ENGINEERING MAINTAINABILITY

Gulf Publishing Company
Houston, Texas

ENGINEERING MAINTAINABILITY

HOW TO

DESIGN FOR

RELIABILITY

AND EASY

MAINTENANCE

B.S. DHILLON

This book is affectionately dedicated to the memory of my grandfather's late brother, Kishan S. Dhillon.

Engineering Maintainability

Gulf Publishing Company
Book Division
P.O. Box 2608, Houston, Texas 77252-2608

10 9 8 7 6 5 4 3 2 1

Library of Congress Cataloging-in-Publication Data

ISBN   0-88415-257-X

Printed in the United States of America.
Printed on acid-free paper (∞).

Contents

Preface

In today's global economy, only those nations that lead in technology will lead the world. Within technology, we include research, development, and manufacturing, as well as maintenance and maintainability. Various studies have indicated that for many large systems or pieces of equipment, maintenance and support account for as much as 60 to 75 percent of their overall life cycle costs, and sometimes much more.

This is probably why there is an increasing emphasis on maintainability during product design. But even though the first book on maintainability, *Electronic Maintainability,* appeared in 1960, there are still only a handful of books available on the subject. In recent years, many new concepts and techniques have been developed to meet the challenges of maintaining modern engineering systems. A professional needing information on these developments generally faces a great deal of difficulty and inconvenience, because they are generally discussed in various technical papers and in specialized books but have not been treated within the framework of a single volume on maintainability. This book is an attempt to present both traditional and modern maintainability concepts in a single volume to meet the challenges of modern system design. Every effort was made to treat the topics discussed in such a manner that the reader will need minimum previous knowledge to understand the contents.

If the reader desires to delve deeper into a specific area, references provide the sources of most of the material presented. Furthermore, the volume contains numerous examples to facilitate understanding of the material, as well as challenging problems to test reader comprehension. Professional engineers, design and maintenance managers, university professors, along with undergraduate and graduate engineering students, should all find this book useful.

I am indebted to my friends, colleagues, students, various maintainability professionals, and others for their interest and encouragement throughout this project. I am grateful to Josee Rocheleau for typing the first draft of this book. I also thank my children, Jasmine and Mark, for their patience during this project. Last, but not least, I thank my wife, Rosy, for typing various portions of this book, for her help in proofreading, and for her unending patience.

B. S. Dhillon, Ph.D.
University of Ottawa

Introduction

WHAT IS MAINTAINABILITY?

Maintainability refers to the measures taken during the development, design, and installation of a manufactured product that reduce required maintenance, manhours, tools, logistic cost, skill levels, and facilities, and ensure that the product meets the requirements for its intended use.

The precise origin of maintainability as an identifiable discipline is somewhat obscured, but in some ways the concept goes back to the very beginning of the twentieth century. For example, in 1901 the Army Signal Corps contract for development of the Wright Brothers' airplane stated that the aircraft should be "simple to operate and maintain" [1]. In the modern context, the beginning of the discipline of maintainability may be traced to the period between World War II and the early 1950s, when various studies conducted by the United States Department of Defense produced startling results [2, 3]:

- A Navy study reported that during maneuvers electronic equipment was operative only 30% of the time.
- The U.S. Army's Eighth Air Force, stationed in Britain during World War II, reported that only 30% of the heavy bombers stationed at one airfield were in operational readiness condition at any given time and that the situation at other airfields was quite similar.
- A study conducted by the Army reported that approximately two-thirds to three-fourths of equipment was either out of service or under repair at any given time.

Another step in the discipline's development was a 12-part series of articles that appeared in *Machine Design* in 1956 and covered subjects such as design of electronic equipment for maintainability, design of maintenance controls, factors to consider in designing displays, design of covers and cases, designing for installation, recommendations for designing maintenance access in electronic equipment, a systematic approach to preparing maintenance procedures, and design recommendations for test points [4]. In 1957, the Advisory Group on Reliability of Electronic Equipment (AGREE), established by the Department of Defense, published a report containing recommendations compiled by nine AGREE task groups. These recommendations have served as a basis for most of the current standards on maintainability [5].

In 1960, the United States Air Force initiated a program to develop an effective systems approach to maintainability, which led to the development of specification MIL-M-26512. This specification provided program guidance and established maintainability procedures to assure that systems and equipment would satisfy qualitative and quantitative operational requirements.

An appendix to specification MIL-M-26512 contained a method for planning maintainability demonstration tests, addressing issues such as sample size, data evaluation, control of the test, and sample selection. It also served as the basis for MIL-STD-471, an important later Department of Defense document on maintainability testing [4, 6].

In the latter part of the 1960s many military documents related to maintainability appeared: MIL-STD-470 (*Maintainability Program Requirements*) [7], MIL-STD-471 (*Maintainability Demonstration*) [8], MIL-HDBK-472 (*Maintainability Prediction*) [9], and MIL-STD-721B (*Definition of Effectiveness Terms for Reliability, Maintainability, Human Factors, and Safety*) [10]. Two important military documents that appeared in the 1970s were AMCP 706-134 (*Engineering Design Handbook: Maintainability Guide for Design*) [11] and AMCP 706-133 (*Engineering Design Handbook: Maintainability Engineering Theory and Practice*) [1].

In 1984, the United States Air Force launched the Reliability and Maintainability (R & M) 2000 Initiative. The main objective of this initiative was to give R & M characteristics equal weight with factors such as performance, cost, and schedule during the development and decision process. The Army and Navy launched similar initiatives. In particular, the Army set a goal of cutting the operating and support costs related to reliability, maintainability, availability, and durability in half by 1991 [5].

Three important commercially available publications that appeared in 1960, 1964, and 1969, respectively were *Electronic Maintainability* [12], *Maintainability* [13], and *Maintainability Principles and Practices* [14]. There have been many important publications since then; references 15 and 16 give a comprehensive list of publications on maintainability.

THE IMPORTANCE, PURPOSE, AND RESULTS OF MAINTAINABILITY EFFORTS

The alarmingly high operating and support costs of systems and equipment, in part due to failures and the necessary subsequent repairs, are the prime reasons for emphasizing maintainability. One study conducted by the United States Air Force in the 1950s reported that one third of all its personnel was involved in maintenance and one third of all Air Force operating costs stemmed from maintenance. Some examples of the costs are the expense of maintenance personnel and their training, maintenance instructions and data, repair parts, test and support equipment, training equipment, repair parts, and maintenance facilities.

The objectives of applying maintainability engineering principles to engineering systems and equipment include:

- Reducing projected maintenance time and costs through design modifications directed at maintenance simplifications
- Determining labor-hours and other related resources required to carry out the projected maintenance
- Using maintainability data to estimate item availability or unavailability

When maintainability engineering principles have been applied effectively to any product, the following results can be expected [1]:

- Reduced downtime for the product and consequently an increase in its operational readiness or availability
- Efficient restoration of the product's operating condition when random failures are the cause of downtime
- Maximizing operational readiness by eliminating those failures that are caused by age or wear-out

MAINTAINABILITY IN THE GOVERNMENT PROCUREMENT PROCESS AND IN THE COMMERCIAL SECTOR

Over the years, the United States Department of Defense (DOD) has played an instrumental role in providing direction and standards for the application of maintainability principles. Even though other government agencies such as the National Aeronautics and Space Administration and the Department of Energy have recognized the importance of maintainability to their missions and have taken appropriate measures, the DOD has been the most effective in implementing maintainability principles [5].

DOD Directive 5000.40 (*Reliability and Maintainability*) outlines specific objectives, policies, and responsibilities for addressing maintainability issues in the procurement process. Specifically, the directive states the need to decrease the demand for maintenance and logistics support and to manufacture items that can be operated and maintained with available manpower, skills, and training facilities in the field. The DOD has produced some of the other key documents on maintainability and has taken various initiatives to improve the reliability and maintainability of its equipment and systems.

Market pressure has been the primary force behind the application of maintainability principles in the commercial sector. But the private sector's emphasis on maintainability is still a long way behind that of the government. Probably the most effective maintainability effort has been in the commercial aircraft industry, where aircraft availability has become an important index of an airline's ability to satisfy the needs of its market. The maintainability of an aircraft has a fundamental influence on its availability or dispatch reliability. Thus the main aim of maintainability efforts is to improve dispatch availability or reliability through factors such as the following [5]:

- **Interchangeability.** This is the extent to which one item can be readily replaced with an identical item without a need for recalibration. Such flexibility in design reduces maintenance work and in turn maintenance costs.
- **Accessibility.** This is the ease and rapidity with which an aircraft part can be reached and the required maintenance performed. Poor accessibility leads to increased downtime and, in turn, lower revenue.

Thus, one design goal should be to provide access to failed parts that does not require removing other parts.

- **Maintenance frequency.** This is the frequency with which each maintenance action must be conducted and is central to the preventive, schedule, or corrective maintenance requirements of an aircraft.
- **Simplicity.** This is the simplification of maintenance tasks associated with the aircraft system. System simplification helps to reduce the costs of spares and improves the effectiveness of maintenance troubleshooting.
- **Visibility.** This measures how readily the aircraft part requiring maintenance can be seen. A blocked view can significantly increase downtime.
- **Testability.** This is the measure of fault detection and fault isolation ability. Fault diagnosis speed can significantly influence downtime and maintenance costs.
- **State-of-the-art.** Technological advances can help to improve maintainability and decrease maintenance costs.

MAINTENANCE ENGINEERING VERSUS MAINTAINABILITY ENGINEERING

Maintenance and maintainability are closely interrelated, and many people find it difficult to make a clear distinction between them. Maintenance refers to the measures taken by the users of a product to keep it in operable condition or repair it to operable condition. Maintainability refers to the measures taken during the design and development of a product to include features that will increase ease of maintenance and will ensure that when used in the field the product will have minimum downtime and life-cycle support costs [1, 17].

More simply [18]:

- Maintenance is the act of repairing or servicing equipment.
- Maintainability is a design parameter intended to minimize repair time.

Maintenance engineers use time design reviews and test results to reduce maintenance support requirements. Because it is the responsibility of the maintenance engineers to ensure that equipment design and development requirements reflect the user's maintenance needs, they are concerned with factors such as system mission, operational, and support profiles, the levels and kinds of maintenance and other support resources required, and the environment in which the system will be operated and maintained.

Because engineers should consider maintenance requirements before designing a product, maintainability design requirements can be determined by processes such as maintenance engineering analysis, the analysis of maintenance tasks and requirements, the development of maintenance concepts, and the determination of maintenance resource needs.

MAINTAINABILITY SCIENCE AND DOWNTIME

Maintainability engineers must use a scientific approach to measure the maintainability qualities they build into manufactured products. They should also use a scientific method to rate and examine indistinct maintainability concepts and to review new ways to rectify deficiencies in products [11]. A scientific approach to maintainability should include good maintainability principles and guidelines; controlling mechanisms to ensure that maintainability is built in; simple and straightforward maintainability measuring methods that mean fewer and less complicated tasks during the design process and other processes; good interaction between design and maintainability professionals; and good record keeping to allow statistical analyses and reviews of maintainability action effectiveness.

Because equipment downtime consists of many components and subcomponents, as shown in Figure 1-1 [19], there are numerous engineering and analytical efforts required to reduce downtime. The three main components of equipment downtime are logistic time, administrative time, and active repair time [20].

Logistic time is that portion of equipment downtime during which repair work is delayed because a replacement part of other component of the equipment is not immediately available.

Apart from the fact that replacement parts needs are affected by factors such as operating conditions and the equipment's inherent capability to tolerate operating stress levels, logistic time is largely a matter of management. Acquisition personnel can play an important role in minimizing logistic time by developing effective procurement policies.

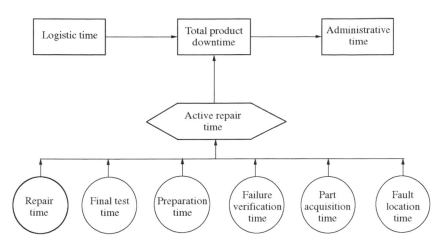

Figure 1-1. Equipment downtime components.

Active repair time is that portion of equipment downtime during which the repair staff is actively working to effect a repair. Its six elements are fault location time, preparation time, failure verification time, actual repair time, part acquisition time, and final test time, as shown in Figure 1-1. Usually, the length of active repair time reflects factors such as product complexity, diagnostic adequacy, nature of product design and installation, and the skill and training of the maintenance staff. All in all, since active repair time is largely determined by the equipment's built-in maintainability characteristics, the manufacturer of the equipment plays a key role in influencing how much active repair time is required.

Administrative time is that portion of equipment downtime not taken into consideration in active repair time and in logistic time. Required administrative activities and unnecessarily wasted time are both included in administrative time. This time is a function of the structure of the operational organization and is influenced by factors such as work schedules and the nontechnical duties of maintenance people.

MAINTAINABILITY STANDARDS, HANDBOOKS, AND INFORMATION SOURCES

Over the years, there have been many standards, specifications, and handbooks developed that directly or indirectly relate to maintainability. Some of these documents are [15, 21, 22]:

- MIL-STD-470A, *Maintainability Program for systems and Equipment,* Department of Defense, Washington, D.C.
- MIL-STD-2084 (AS), *General Requirements for Maintainability,* Department of Defense, Washington, D.C.
- MIL-STD-2165, *Testability Program for Systems and Equipment,* Department of Defense, Washington, D.C.
- MIL-STD-471A, *Maintainability/Verification/Demonstration/Evaluation,* Department of Defense, Washington, D.C.
- MIL-HDBK-472, *Maintainability Prediction,* Department of Defense, Washington, D.C.
- MIL-STD-721C, *Definition of Terms for Reliability and Maintainability,* Department of Defense, Washington, D.C.
- AMCP 706-134, *Maintainability Guide for Design,* Department of Defense, Washington, D.C.
- AMCP 706-133, *Maintainability Engineering Theory and Practice,* Department of Defense, Washington, D.C.
- *Handbook of Product Maintainability,* Reliability Division, American Society for Quality Control, Milwaukee, Wisconsin, 1973.
- ISO 8107, *Nuclear Power Plant Maintainability Terminology,* International Organization for Standardization, Geneva, Switzerland.
- MOD UK DSTAN 00-25: *Part II, Human Factors for Designers of Equipment; Part II: Design for Maintainability, Issue 1* (08.88), Department of Defense, London, U.K.
- NATO ARMP-1 ED2, *NATO Requirements for Reliability and Maintainability,* North Atlantic Treaty Organization, Brussels, Belgium.
- NATO ARMP-5 AMDO, *Guidance on Reliability and Maintainability Training,* North Atlantic Treaty Organization, Brussels, Belgium.
- NATO ARMP-8 AMDO, *Reliability and Maintainability in the Procurement of Off-The-Shelf Equipment,* North Atlantic Treaty Organization, Brussels, Belgium.
- NATO STANAG 4174 ED 1 AMD1, *Allied Reliability and Maintainability Publications,* North Atlantic Treaty Organization, Brussels, Belgium.

Other sources of maintainability-related information are:

- Ankenbrandt, F. L., editor. *Electronic Maintainability.* Engineering Publishers, Elizabeth, New Jersey, 1960.
- Goldman, A. S. and Slattery, T. B. *Maintainability.* John Wiley and Sons, New York, 1964.

- Blanchard, B. S. and Lowery, E. E. *Maintainability Principles and Practices.* McGraw-Hill Book Company, New York, 1969.
- Cunningham, C. E. and Cox, W. *Applied Maintainability Engineering.* John Wiley and Sons, New York, 1972.
- Smith, D. J. and Babb, A. H. *Maintainability Engineering.* Pitman, New York, 1973.
- Blanchard, B. S., Verma, D. and Peterson, E. L. *Maintainability: A Key to Effective Serviceability and Maintenance Management.* John Wiley and Sons, New York, 1995.

Maintainability Data Sources

- IEC 706 PT3, *Guide on Maintainability of Equipment, Part III: Sections Six and Seven, Verification and Collection, Analysis and Presentation of Data,* first edition, International Electrotechnical Commission, Geneva, Switzerland.
- RAC EEMD1, *Electronic Equipment Maintainability Data.* Reliability Analysis Center, Rome Air Development Center, Griffis Air Force Base, Rome, New York.
- SAA AS2529, *Collection of Reliability, Availability and Maintainability Data for Electronics and Similar Engineering Use* (R 1994), Standards Association of Australia, Melbourne.
- GIDEP Data. The Government Industry Data Exchange Program (GIDEP) is a computerized data bank. Originally, GIDEP was jointly established by the National Aeronautics and Space Administration, the Canadian Military Electronics Standards Agency, the U.S. Air Force Logistics Command, the U.S. Navy, the U.S. Army, and the U.S. Air Force Systems Command. The data bank is managed by the GIDEP Operations Center, Fleet Missile Systems, Analysis and Evaluation, Department of Defense, Corona, California.
- *Reliability and Maintainability Data for TD-84-3, Industrial Plants,* A.P. Harris and Associates, Ottawa, Ontario, 1984.

MAINTAINABILITY TERMS AND DEFINITIONS

Some of the terms and definitions used in maintainability work are [1, 20, 23, 24]:

- **Maintainability.** This refers to the aspects of a product that increase its serviceability and repairability, increase the cost-effectiveness of maintenance, and ensure that the product meets the requirements for its intended use.
- **Downtime.** This is the total time during which the product is not in an adequate operating state.
- **Repairability.** This is the probability that a failed product will be repaired to its operational state within a given active repair time.
- **Serviceability.** This is the degree of difficulty or ease with which a product can be restored to its operable state.
- **Availability.** This is the probability that a product is available for use when needed.
- **Active repair time.** This is that segment of downtime during which repair staff work to effect a repair.
- **Logistic time.** This is that segment of downtime occupied by the wait for a needed part or tool.
- **Design adequacy.** This is the probability that the product will complete its intended mission successfully when it is used according to its design specifications.

PROBLEMS

1. Write an essay on the history of maintainability.
2. What are the reasons for the emphasis on maintainability during design?
3. Discuss the results that can be expected when maintainability principles are applied during the product design.
4. Define the following terms:
   - Repairability
   - Maintainability
   - Active repair time
5. Describe the following:
   - Maintainability in the government procurement process
   - Maintainability in the commercial sector
6. Compare between maintenance and maintainability.
7. Discuss the principal elements of active repair time.
8. Discuss at least two data banks for obtaining maintainability-related information.
9. Discuss the principal components of downtime.

REFERENCES

1. AMCP 706-133, *Engineering Design Handbook: Maintainability Engineering Theory and Practice.* Department of Defense, Washington, D.C., 1976.
2. Moss, M. A. *Minimal Maintenance Expense.* Marcel Dekker, Inc., New York, 1985.
3. Shooman, M. L. *Probabilistic Reliability: An Engineering Approach.* McGraw-Hill Book Company, New York, 1968.
4. Retterer, B. L. and Kowalski, R. A. "Maintainability: A Historical Perspective." *IEEE Transactions on Reliability,* Vol. 33, April 1984, pp. 56–61.
5. SAE G-11, *Reliability, Maintainability, and Supportability Guidebook.* The Society of Automotive Engineers, Warrendale, Pennsylvania, 1990.
6. Retterer, B. L. and Griswold, G. H. "Maintenance Time Specification." *Proceedings of the Tenth National Symposium on Reliability and Quality Control,* 1963, pp. 1–9.
7. MIL-STD-470, *Maintainability Program Requirements.* Department of Defense, Washington, D.C., 1966.
8. MIL-STD-471, *Maintainability Demonstration.* Department of Defense, Washington, D.C., 1966.
9. MIL-HDBK-472, *Maintainability Prediction.* Department of Defense, Washington, D.C., 1966.
10. MIL-STD-721B, *Definition of Effectiveness Terms for Reliability, Maintainability, Human Factors, and Safety.* Department of Defense, Washington, D.C., 1966.
11. AMCP 706-134, *Engineering Design Handbook: Maintainability Guide for Design.* Department of Defense, Washington, D.C., 1972.
12. Akenbrandt, F. L., editor. *Electronic Maintainability.* Engineering Publishers, Elizabeth, New Jersey, 1960.
13. Goldman, A. S. and Slattery, T. B. *Maintainability.* John Wiley and Sons, New York, 1964.
14. Blanchard, S. S. and Lowery, E. E. *Maintainability Principles and Practices.* McGraw-Hill Book Company, New York, 1969.
15. Dhillon, B. S. *Reliability Engineering in Systems Design and Operation.* Van Nostrand Reinhold Company, New York, 1983.
16. Dhillon, B. S., *Reliability and Quality Control: Bibliography on General and Specialized Areas.* Beta Publishers, Inc., Gloucester, Canada, 1993.
17. Downs, W. R. "Maintainability Analysis Versus Maintenance Analysis: Interfaces and Discrimination," *Proceedings of the Annual Reliability and Maintainability Symposium,* 1976, pp. 476–481.

18. Smith, D. J. and Babb, A. H. *Maintainability Engineering.* John Wiley and Sons, Inc., New York, 1973.
19. Pecht, M. editor. *Product Reliability, Maintainability, and Supportability Handbook.* CRC Press, Inc., Boca Raton, Florida, 1995.
20. Von Alven, W. H., editor. *Reliability Engineering.* Prentice-Hall, Inc., Englewood Cliffs, New Jersey, 1964.
21. Wilbur, J. W. and Fuqua, N. B. *A Primer for DOD Reliability, Maintainability, and Safety Standards.* The Reliability Analysis Center, Rome Air Development Center, Griffis Air Force Base, New York, 1988.
22. "List of Quality Standards, Specifications, and Related Documents." *Quality Progress,* September 1976, pp. 30–35.
23. Naresky, J. J. "Reliability Definitions." *IEEE Transactions on Reliability,* Vol. 19, 1970, pp. 198–200.
24. MIL-STD-721C, *Definition of Terms for Reliability and Maintainability.* Department of Defense, Washington, D.C.

Maintainability Management

INTRODUCTION

The effective practice of maintainability design and engineering requires a systematic management approach. Maintainability management may be discussed from a variety of perspectives. Among them are the management of maintainability as an engineering discipline, the place of the maintainability function within an organization's structure, and the role maintainability plays at each phase in the life cycle of the product being developed. That is, there are aspects of maintainability management and organization that may need to be altered depending upon the current design stage or phase in the life cycle. Effective maintainability management depends upon other factors as well, including [1]:

- Acceptance by upper management that maintainability is an important characteristic of equipment design
- Establishment of a maintainability engineering entity at a level within the organization that allows effective relationships and functions with respect to other organization entities [2]
- Acceptance throughout the organization of maintainability as a technical discipline on a par with design, maintenance, human factors, reliability, testing and evaluation, integrated logistic

support, and safety engineering, which are all disciplines to which
maintainability has strong links
• Planning, organizing, directing, controlling, budgeting, and monitoring of the maintainability function in a manner similar to the
management of other disciplines

This chapter describes some of the most important aspects of
maintainability management. For a more detailed discussion, see B. S.
Dhillon and H. Reiche, *Reliability and Maintainability Management* [3].

MAINTAINABILITY MANAGEMENT FUNCTIONS IN THE PRODUCT LIFE CYCLE

An important element in achieving an efficient and effective design
is serious consideration of the maintainability issues that arise throughout the product life cycle. An effective maintainability program incorporates a dialogue between the user and manufacturer during the total
life cycle of the product. This dialogue concerns the user's maintenance requirements and other requirements for the product and the
manufacturer's response to those requirements.

The product life cycle is composed of the four phases shown in
Figure 2-1: the concept development phase, the validation phase, the
production phase, and the operation phase. Specific maintainability
functions are associated with each of these phases.

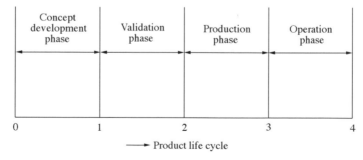

Figure 2-1. Product life cycle phases.

Concept Development Phase

During this phase, the operation needs of the product are translated into a set of operational requirements, and high-risk areas are identified. In other words, the objective of the concept development phase is to develop and choose the most appropriate method of meeting the identified operational needs. The method must be proven viable from technical, schedule, and cost standpoints. A product development plan, implementation plans for the recommended method, advanced development objectives, and any other necessary plans or objectives should also be prepared.

The primary maintainability task during this phase is to determine product effectiveness requirements and to determine, from the purpose and intended operation of the product, the field support policies and other provisions required. Product effectiveness can be defined as the probability that the product can successfully satisfy an operational demand within a defined interval when used according to design specifications. The most frequently used measures of product effectiveness are probabilities, expected value, and rates [1].

In order to establish product maintainability requirements, it is necessary to determine product utilization rates, mission time factors, and product life cycle duration, including product use and out-of-service conditions. It is also necessary to describe mission and performance expectations, product operating modes, and the overall logistic support objectives and concepts.

Validation Phase

In this phase the operational requirements developed during the concept development phase are refined further in terms of product design requirements. The main objective of the validation phase is to ensure that no full-scale development takes place until associated costs, schedules, and performance and support objectives have been prepared and evaluated with utmost care.

During this phase, maintainability management tasks include:

- Developing a maintainability program plan that meets contractual requirements
- Developing a plan for maintainability testing and demonstration

- Determining specific maintainability, reliability, and product effectiveness requirements
- Developing maintainability incentives or penalties
- Coordinating and monitoring maintainability efforts throughout the company
- Developing maintainability policies and procedures for both the validation phase and the subsequent full-scale engineering effort
- Providing assistance to maintenance engineering in areas such as performing maintenance analysis and developing logistic policies and product effectiveness requirements
- Developing a planning document for data collection, analysis, and evaluation
- Performing maintainability predictions and allocations and participating in trade-off analyses
- Participating in design reviews

Because the most crucial part of the maintainability effort occurs during the concept development and validation phases, effective maintainability management is especially important at these stages.

In particular, some of the functions that must be completed before the production phase are updating the maintainability program plan to meet final specification requirements for the project; monitoring the maintainability efforts of subcontractors; issuing detailed program schedules, milestones, work orders, and budgets, and periodically reviewing and updating them; predicting and addressing maintainability requirements in quantitative terms, down to the lowest-level product component; preparing specific maintainability test and demonstration plans; and monitoring the maintainability effort, following the maintainability program plan and management policies and procedures.

Production Phase

During this phase the product is manufactured, tested, delivered, and in some cases installed in accordance with the technical data developed in the earlier phases. Even though at this stage the maintainability design effort will be largely completed, design should be reviewed and updated as engineering changes, initial field experience, and logistic support modification require. The production phase maintainability effort includes production process monitoring; examining production test trends from the standpoint of adverse effects on maintainability and maintenance requirements; evaluating all proposals for change with respect to their impact on maintainability; ensuring the eradication of all discrepancies that may diminish maintainability; and taking part in the development of controls for process variations, errors, and other problems that may affect maintainability.

Operation Phase

During this phase, the user puts the product into operation, logistically supports it, and modifies it as appropriate. It is in this phase that the supply, maintenance, training, overhaul, and material readiness requirements and characteristics of the product become clear. Therefore, although there are no specific maintainability requirements at this time, the phase is probably the most significant because the product's true cost-effectiveness and logistic support are now demonstrated and maintainability data can be collected from the experience for use in future applications.

MAINTAINABILITY ORGANIZATION FUNCTIONS AND TASKS

The functions and tasks performed by the maintainability organization may be grouped into many general functions: administrative, analysis, design, documentation, and coordination [1].

Administrative

The administrative function encompasses those tasks concerned with cost, performance, and schedule, and it provides overall direction and control to maintainability program management [4]. Some of the tasks involved are preparing a maintainability program plan; organizing the maintainability effort; participating in design reviews and program management; assigning maintainability-related responsibilities, tasks, and work orders; preparing budgets and schedules; developing and issuing policies and procedures to be used in maintainability efforts; participating in meetings and conferences regarding maintainability management; monitoring the maintainability organization's output; acting as a liaison with upper-level management and other concerned bodies; and providing maintainability training as appropriate.

Analysis

A maintainability organization spends considerable effort on various analytical projects such as maintainability allocation, maintainability prediction, and field data evaluation. Some tasks related to maintainability analysis are examining product specification documents with respect to maintainability requirements; performing maintainability allocation and prediction studies; analyzing maintainability feedback data obtained from the field and other sources; taking part in product engineering analysis to safeguard maintainability interests; preparing maintainability demonstration documents; participating in or performing required maintenance analysis; and participating in meetings and conferences regarding maintainability analysis.

Design

Product maintainability design deals with those features and characteristics of the product that will increase ease of maintenance, make maintenance more cost-effective, and in turn lower logistic support needs. Some of the activities involved are reviewing product design with respect to maintainability features, preparing maintainability design documents, taking part in the development of maintainability design criteria and guidelines, participating in design reviews to safeguard the interests of maintainability, approving design drawings from the standpoint of maintainability, participating in meetings and conferences regarding maintainability, and providing consulting services to professionals such as design engineers.

Documentation

The maintainability effort produces and uses a significant amount of information and data. For the sake of achieving a cost-effective, coherent, and comprehensive design, the effective and efficient handling of this information is crucial. Maintainability documentation consists of tasks such as developing maintainability data and feedback reports, establishing and maintaining a maintainability data bank and a library containing important maintainability documents and information, documenting information related to maintainability management, preparing and maintaining handbook data and information with respect to maintainability, documenting maintainability trade-offs and the results of maintainability analysis, and documenting the results of maintainability design reviews.

Coordination

Coordination and liaison account for much of the maintainability management effort. This coordination effort is a critical factor in assuring an effective and optimized design. Some of the elements of maintainability coordination are interfacing with product engineering and other engineering disciplines; acting as a liaison with subcontractors on maintainability matters; coordinating with bodies such as governments, professional societies, and trade associations on maintainability-related activities; and coordinating maintainability training and information efforts for all personnel involved.

MAINTAINABILITY ORGANIZATIONAL STRUCTURES

For an effective maintainability effort, it is important to carefully decide where to place the maintainability organization within the overall organizational structure. There is no one conventionally accepted norm. Manufacturers and user organizations have structured the maintainability function in many different ways, depending on factors such as the enterprise's organizational philosophy and its method of doing business, the overall size of the enterprise, the size and complexity of projects, and the emphasis placed on maintainability. Nonetheless, top-level management should carefully consider the following ways the maintainability function can fit into the enterprise's overall structure [1, 5]:

- As a fully integrated, undifferentiated element of the engineering organization

- As a centralized, distinct line organization within the framework of the engineering department
- As a staff function that operates in an advisory capacity to project management and in a consultative capacity to designers
- As a decentralized element of program management or system engineering in an organization with a project or matrix structure

In small enterprises, maintainability tends to be a fully integrated part of the engineering design team's work. The following treats the other kinds of structures, in which one or more distinct maintainability groups exist.

Centralized Organization

As shown in Figure 2-2, in this case the maintainability function operates as a distinct line organization within the overall engineering department. The entire maintainability effort is centralized under the authority of a single manager or chief. This arrangement gives emphasis to maintainability as a design discipline and is most effective and efficient when both managers and engineers recognize maintainability as a natural part of good engineering design. It works particularly well when there is only one major project or there are a number of small projects involving basically similar products or customers.

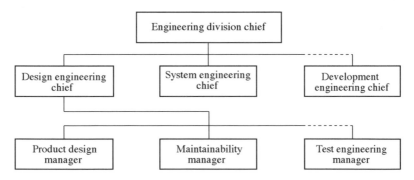

Figure 2-2. Maintainability as a line organization within the overall engineering department.

Advisory or Consultative Staff Function

In this case, maintainability is considered to be basically a management and analytic function rather than a design function. Maintainability personnel provide services to system engineers such as performing maintainability demonstration requirements, computing mean time to repair, and analyzing maintenance tasks and requirements with regard to product effectiveness and product design specifications. They also provide consulting services to design engineers. In many enterprises that use this structure, the maintainability staff reports to the system effectiveness manager, who in turn is under the authority of a system engineering manager.

Decentralized Organization

Establishments that regularly handle large and complex projects often break down the maintainability effort according to the functions and tasks described earlier. In one example of such a structure, a small maintainability program group, located in the project office, handles maintainability-related program and coordination activities. The second maintainability group performs analytic activities within the product effectiveness organization. The third maintainability group, assigned to the design engineering organization, deals with maintainability design features. The fourth maintainability group falls within the overall documentation and data organization and handles maintainability documentation requirements. This type of arrangement is often found in aerospace and military organizations that have a project or matrix organizational structure.

It can be difficult to coordinate the efforts of different maintainability groups separated from one another on the management ladder. There is therefore the risk of decreased efficiency with this kind of structure.

MAINTAINABILITY PROGRAM PLAN

During the conceptual design phase, the overall system requirements are established on the basis of customer needs and maintainability planning begins. The latter means the preparation of the maintainability program plan. Normally, the person who is responsible for implementing program requirements also takes responsibility for preparing the plan. But depending on the nature of the project under consideration, either the user or the manufacturer could develop the maintainability program plan. Figure 2-3 shows some of the key elements of a maintainability program plan are [5]: objectives; references; technical communications; organization; maintenance concept; maintainability design criteria; policies and procedures; organizational interfaces; maintainability program tasks; program review, evaluation, and control; and subcontractor/supplier activity.

The objective is a description of the overall requirements for the maintainability program and the plan's goals. It also usually includes a brief description of the product under consideration. The reference section lists all documents relevant to the maintainability program requirements, such as specifications, applicable standards, and plans. The technical communications section briefly describes each deliverable data item and the associated due dates. The organization section depicts the overall structure of the enterprise and provides a detailed organizational breakdown of the maintainability group involved in the project. This section also shows the work breakdown structure, describes the background and experience of the maintainability group personnel, and indicates the personnel assigned to each task.

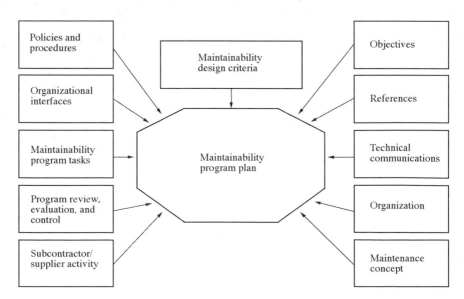

Figure 2-3. Maintainability program plan elements.

The maintenance concept section presents the product's basic maintenance requirements and describes in sufficient depth issues such as qualitative and quantitative objectives for maintenance and maintainability, spare and repair part factors, operational and support concepts, test and support equipment criteria, and organizational responsibilities. Similarly, the maintainability design criteria describes or references specific maintainability design features applicable directly to the product under consideration. Also, this description may relate to quantitative and qualitative factors concerning part selection, interchangeability, accessibility, or packaging. The main purpose of the policies and procedures section is to assure customers that the group implementing the maintainability program will do so effectively. The section also references or incorporates management's overall policy directives for maintainability efforts. The directives address topics such as techniques to be used for maintainability allocation and prediction, task analysis, maintainability demonstration methods, participation in design review and evaluation, data collection and analysis, and trade-off studies.

The organizational interfaces section discusses the relationships and lines of communication between the maintainability organization and the overall organization. Areas of interface include product engineering, reliability engineering, testing and evaluation, design, human factors, and logistic support, as well as customers and suppliers. The maintainability program tasks, or work statement, section describes each program task, task schedule and major milestones, task input requirements, expected task output results, and projected cost. The program review, evaluation, and control section describes the methods to be used for program reviews, technical design review, and feedback and control. The section also discusses the evaluation and incorporation of proposed changes and appropriate corrective measures to be initiated in given situations, and outlines a risk management plan. A final section addresses the organization's relationships with subcontractors and suppliers connected to the maintainability program, and procedures to be used for review and control within those relationships.

PERSONNEL ASSOCIATED WITH MAINTAINABILITY

Beside maintainability engineers, there are many other engineering professionals who indirectly contribute to product maintainability. Among them are reliability engineers, quality control engineers, human factors engineers, and safety engineers. This section discusses the functions of maintainability engineers and of these other professionals [3, 6].

Maintainability Engineer

This person plays an instrumental role in product design, and has often served before as a design engineer. Some of a maintainability engineer's tasks are maintainability prediction, maintainability allocation, developing maintainability demonstration documents, performing analysis of maintainability feedback data, reviewing product design with respect to maintainability features, participating in the development of maintainability design criteria, and preparing maintainability design documents.

Reliability Engineer

As is the case with the maintainability engineer, the reliability engineer is often a former design engineer. This person assists management in defining, evaluating, and containing risks. Responsibilities include reliability allocation, reliability prediction, analyzing customer requirements, developing reliability growth monitoring procedures and reliability evaluation models, monitoring subcontractor reliability programs, participating in design reviews, developing reliability test and demonstration procedures, and assessing the effect of environment on product reliability.

Quality Control Engineer

While this individual does not necessarily belong to one of the engineering disciplines, he or she plays an important role during the product life cycle, especially during the design and manufacture phases. The quality control engineer develops and applies appropriate inspection plans, evaluates the quality of procured parts, develops quality related standards, analyzes quality associated defects, uses statistical quality control methods, participates in quality related meetings, and audits the quality control system at appropriate times [7].

Human Factors Engineer

The maintainability of the product will depend to a large degree on the decisions made by the human factors engineer, who should be trained in fields such as anthropometry, physiology, and/or psychology [8]. The human factors engineer determines the tasks required to make full use of the product; the most appropriate way of grouping the various tasks required into individual jobs; the best way of displaying specified information, and of arranging visual displays to ensure optimum use, especially when the user must split his or her attention between two or more displays; how to label control devices most clearly; the arrangement of control devices that generates optimum use; what level of information flow or required decision making over-burdens operators; how human decision-making and adaptive abilities can be put to best use; the environmental conditions that affect individual performance and physical well-being; and other similar matters [9].

Safety Engineer

The role played by this professional during the design phase is very critical, as the newly designed product must be safe to operate and maintain. Some of the safety engineer's tasks include analyzing historical data on product hazards, failures, and accidents; analyzing new designs from the standpoint of safety; keeping management continually informed about safety program performance; monitoring subcontractors' safety efforts; providing safety-related information to those concerned; establishing criteria for analyzing any accident associated with a product manufactured by the company; and developing safety warning devices.

MAINTAINABILITY DESIGN REVIEWS

The design review comes during the product design phase and is a critical component of modern design practice; its primary objective is to determine the progress of the design effort and to ensure that correct design principles are being applied. Participants in a design review evaluate items such as sketches, models, drawings, assemblies, and mock-ups associated with a product. They assess potential and existing problems related to manufacture of the product, its functional capability, logistics support it requires, and its safety, reliability, maintainability, and human factors aspects [10]. Usually, representatives of the various groups responsible for these aspects of the product attend design reviews, as the decisions made during these reviews determine future change to the design. According to some experts, design reviews represent between 1 and 2% of the overall engineering cost of a product [11].

Design reviews fall into three distinct categories: the preliminary design review, the intermediate design review, and the final design review. All three reviews require input as applicable from maintainability staff.

The preliminary design review comes before formulation of the initial design. Its purpose is the careful examination of the functions the product must perform and the standards it must meet; the accuracy, validity, and completeness of the design specification requirements that have been established; any design constraints; the available data on similar products; cost objectives; current and future availability of the necessary components or materials; scheduling requirements; and required tests and documentation [11, 12].

The intermediate design review, held before the detailed production drawings are developed, has as its purpose comparison of each specification requirement with the proposed design. These requirements may involve maintainability, maintenance, reliability, safety, human factors, cost, performance, schedule, and usage of standard parts.

The final design review, sometimes also referred to as the critical design review, takes place soon after the completion of production drawings. At this stage, the design team has available information such as cost data, test results, and the results of preceding design reviews. This review emphasizes factors such as manufacturing methods, value engineering, design producibility, analysis results, and quality control of incoming parts.

There are many maintainability issues that require attention during design reviews. Some of these are [10]:

- Design constraints and specified interfaces
- Maintainability prediction results
- Use of on-line repair with redundancy
- Results of maintainability trade-off studies
- Results of failure mode and effect analysis
- Use of unit replacement approach
- Identified maintainability problem areas and proposed corrective measures
- Use of automatic test equipment
- Use of unit-replacement approach

- Use of built-in monitoring and fault-isolation equipment
- Physical configuration and layout drawings, and schematic designs
- Conformance to maintainability design specifications
- Selection of parts and materials
- Verification of maintainability design test plans
- Maintainability assessment using test data
- Assessment of maintenance and supportability
- Maintainability demonstration test data
- Maintainability test data obtained from experimental models and breadboards
- Corrective measures proposed and taken
- Proposed alternatives for correction action
- Proposed changes to the maintenance concept
- Summary of maintainability design status
- Summary of current and potential maintainability problems identified

Design Review Team

The members of the design review team share the overall objective—to produce an effective product or system. The maintainability engineer or specialist is an important member of this team. The team usually consists of one or more project engineers, design engineers, senior engineers, electrical, mechanical, and manufacturing engineers, management representatives, logistics representatives, maintenance representatives, and customer representatives. The number of design review team members may vary from one project to another but should never be more than twelve [11]. The team members raise questions about subjects such as performance; maintainability, including interchangeability of parts, approach to minimizing downtime, and the maintenance philosophy being applied; specifications, including the validity of the specifications that have been established and adherence to those specifications; standardization; safety; value engineering; reliability, including reliability allocation, reliability predictions, reliability tests, and results of failure modes and effect analysis; human factors issues, such as the design, labeling, and marking of controls and displays, and the reduction of glare; electrical issues, among them circuit analysis, design simplification, and prevention of electrical interference; mechanical issues, such as thermal analysis, balance, and connectors; finishing; product reproducibility, including machine tool effectiveness, part suppliers, and assembly economics; and drafting issues, such as dimensions, tolerances, accuracy, and completeness.

Design Review Board Chairman

The design review board chairman is probably the most important member of the design review team and should always belong to the engineering department. Often, especially in the case of military equipment development, the configuration manager leads the design review team [13]. The following factors need to be considered in appointing the design review board chairman:

- He/she should not be in a direct line of authority over the professionals whose work is to be reviewed.
- He/she should effectively understand the technical problem under consideration.
- He/she should possess the necessary level of skill to lead a technical team.
- He/she should possess a pleasant personality.
- He/she should be capable of exercising tact and discretion as the need arises.

The design review board chairman leads the design review meetings, establishes guidelines for selecting items for review, determines the types of design reviews to be performed and their frequency, evaluates the results of each review and directs follow-up actions, distributes the agenda and related material to all concerned parties well ahead of each design review, supervises publication of the minutes of each review and distributes them, and coordinates the team's efforts. Even though a design review board chairman may establish his or her own style of leading design review meetings, it will usually be important to:

- Describe the design approach to be followed.
- Highlight important factors associated with the design under consideration.
- Provide appropriate background information on the design to be reviewed.
- Highlight expected difficulties with the proposed design.
- Discuss how the proposed design satisfies the original specifications.

Maintainability Design Review Checklist

The maintainability staff usually develops a checklist to assist in reviewing design from the maintainability perspective. This checklist should include all important areas related to maintainability in addition to individual features considered crucial to the type of product under review. One important advantage of developing the maintainability checklist is that it can also allow individuals unfamiliar with maintainability to conduct the design review when people with expertise are unavailable. The checklist may encompass many subjects [14, 15]: the maintenance plan, modular versus non-modular decisions, built-in test equipment, standard circuits, simplicity of design, support requirements, criticality of adjustment requirements, adequacy and location of test points, calibration requirements, criticality of design as it affects maintenance, logistic interfaces, ease of maintenance, testing methods and their adequacy, packaging concepts, producibility versus maintainability, failure records versus maintainability, engineering changes versus maintainability, and comparison of maintainability goals and specifications to attained maintainability.

PROBLEMS

1. Discuss maintainability functions in the product life cycle.
2. Discuss the following maintainability organization functions:
   • Design
   • Analysis
   • Administrative
3. What are the options that top-level management should carefully consider when placing the maintainability function within the enterprise's overall structure?
4. Describe three different structures of maintainability in an organizational setup.
5. Discuss the following elements of a maintainability program plan:
   • Maintainability design criteria
   • Organizational interfaces
   • Maintenance concept
   • Maintainability program tasks
   • Subcontractor/supplier activity
   • Policies and procedures

6. Discuss the functions of the following two professionals:
   • Maintainability engineer
   • Reliability engineer
7. Describe the following two design reviews:
   • Preliminary design review
   • Critical design review
8. List the maintainability-related topics normally reviewed during product design reviews.
9. Discuss the composition of a typical design review team.
10. What are the factors to be carefully considered in appointing the design review board chairman?
11. Develop a 20-question maintainability checklist for use in design review meetings.

REFERENCES

1. AMCP-706-133, *Engineering Design Handbook: Maintainability Engineering Theory and Practice.* Department of Defense, Washington, D.C., 1976.
2. Blanchard, B. S. and Lowery, E. E. *Maintainability.* McGraw-Hill Book Company, New York, 1969.
3. Dhillon, B. S. and Reiche, H. *Reliability and Maintainability Management.* Van Nostrand Reinhold Company, New York, 1985.
4. MIL-STD-470, *Maintainability Program Requirements for Systems and Equipment.* Department of Defense, Washington, D.C., 1966.
5. Blanchard, B. S., Verma, D. and Peterson, E. L. *Maintainability: A Key to Effective Serviceability and Maintenance Management.* John Wiley & Sons, New York, 1995.
6. Dhillon, B. S., *Engineering Management: Concepts, Procedures and Models.* Technomic Publishing Co., Inc., Lancaster, Pennsylvania, 1987.
7. Hayes, G. E. and Romig, H. G. *Modern Quality Control.* Collier MacMillan Publishers, London, 1977.
8. Woodson, W. E. *Human Factors Design Handbook.* McGraw-Hill Book Company, New York, 1981.
9. McCormick, E. J. *Human Factors Engineering.* McGraw-Hill Book Company, New York 1970.
10. Patton, J. D. *Maintainability and Maintenance Management.* Instrument Society of America, Research Triangle Park, North Carolina, 1980.

11. Hill, P. H. *The Science of Engineering Design.* Holt, Rinehart, and Winson, New York, 1970.
12. Carter, C. L. *The Control and Assurance of Quality, Reliability, and Safety.* C. L. Carter & Associates, Inc., Richardson, Texas, 1978.
13. AMCP-706-196, *Engineering Design Handbook: Development Guide for Reliability (Design Reliability).* Department of Defense, Washington, D.C., 1976.
14. AMCP-706-134, *Engineering Design Handbook: Maintainability Guide for Design.* Department of Defense, Washington, D.C., 1970.
15. Pecht, M., editor. *Product Reliability, Maintainability and Support-ability Handbook.* CRC Press, Inc., New York, 1995.

Maintainability Measures, Functions, and Models

INTRODUCTION

Reliability and maintainability are important measures of the effectiveness of systems or products. One way to define the difference between reliability and maintainability is that while reliability is the probability that a failure will not occur in a particular time, maintainability is the probability that required maintenance will be successfully completed in a given time period. Maintainability is a design characteristic that affects accuracy, ease, and time requirements of maintenance actions. It may be measured by combining factors such as frequency of maintenance, maintenance costs, elapsed maintenance or repair times, and labor hours.

These measures make possible the quantitative assessment of product maintainability. The primary purpose of maintainability measures is to influence design and subsequently produce a more cost-effective and maintainable end product.

When a repair starts at time t = 0, the probability that the repair will be completed in a given time t will affect a variety of decisions. Many mathematical models are available to calculate associated measures and functions of maintainability.

MAINTAINABILITY MEASURES

The measures used in maintainability analysis include mean time to repair, mean active preventive maintenance time, and mean active corrective maintenance time, maximum corrective maintenance time, and mean maintenance downtime [1–4].

Mean Time to Repair (MTTR)

This widely used maintainability measure is easy to quantify. MTTR measures the elapsed time required to perform a given maintenance activity and is subsequently used to calculate system availability and downtime.

MTTR, also referred to as mean corrective maintenance time or typical corrective maintenance or repair cycle, is composed of five steps: fault/failure detection, fault/failure isolation, disassemble to gain access, repair, and reassembly.

MTTR is defined by

$$\text{MTTR} = \left(\sum_{i=1}^{m} \lambda_i T_i \right) / \sum_{i=1}^{m} \lambda_i \tag{3.1}$$

where m is the total number of units.

T_i is the corrective maintenance or repair time needed to repair unit i; for i = 1, 2, 3, . . . , m.

λ_i is the constant failure rate of unit i; for i = 1, 2, 3, . . . , m.

Exponential, lognormal, and normal probability distributions can all represent mean time to repair. The exponential distribution is assumed for electronic equipment with an effective built-in test capability along with a rapid remove-and-replace maintenance concept. However, the exponential assumption for mean time to repair may lead to wrong conclusions as most repair actions consume some degree of repair time.

The lognormal distribution is often assumed for electronic equipment without a built-in test capability, and it can also be used for electromechanical systems having widely variant individual repair times. The normal distribution is normally assumed for mechanical or electromechanical equipment with a remove-and-replace maintenance concept.

Except for the exponential distribution, MTTR does not provide sufficient information concerning the tails of the distribution, such as the frequency and duration of excessively long maintenance actions [2]. Nonetheless, it is still an important design parameter for complex systems and can be measured by testing the hardware.

Example 3-1

Assume that a system is composed of four replaceable subsystems, A, B, C, and D, with respective failure rates as follows: $\lambda_A = 0.0001$ failures/hour, $\lambda_B = 0.0002$ failures/hour, $\lambda_C = 0.0003$ failures/hour, and $\lambda_D = 0.0004$ failures/hour. The respective corresponding corrective maintenance times for subsystems A, B, C, and D are $T_A = 0.5$ hour, $T_B = 1$ hour, $T_C = 1.5$ hours, and $T_D = 2$ hours. To calculate the mean time to repair (MTTR), we substitute the given data into Equation 3.1 and get

$$\text{MTTR} = \frac{(0.0001)(0.5) + (0.0002)(1) + (0.0003)(1.5) + (0.0004)(2)}{(0.0001) + (0.0002)(0.0003) + (0.0004)}$$

$$= 1.5 \text{ hours}$$

The system mean time to repair is 1.5 hours.

Mean Preventive Maintenance Time

Preventive maintenance activities such as inspections, calibrations, and tuning keep equipment at a specified performance level. The objective of a preventive maintenance program is to postpone the point at which the equipment or any of its components wears out or breaks down. A carefully planned preventive maintenance program can help to reduce the equipment's downtime and improve its performance. On the other hand, a badly tailored preventive maintenance program can have a negative impact on equipment operation.

The mean preventive maintenance time is expressed by

$$T_{mp} = \frac{\sum_{i=1}^{k} (T_{mpi})(F_{pti})}{\sum_{i=1}^{k} F_{pti}} \tag{3.2}$$

where T_{mp} is the mean preventive time.

T_{mpi} is the elapsed time for preventive maintenance task i, for i = 1, 2, 3, . . . , k.

F_{pti} is the frequency of preventive maintenance task i, for i = 1, 2, 3, . . . , k.

k is the number of preventive maintenance tasks.

It is to be noted that if the frequencies F_{pti} are specified in maintenance tasks per hour, then the T_{mpi} should also be expressed in hours.

Median Corrective Maintenance Time

This is a measure of the time within which 50% of all corrective maintenance can be completed. Calculation of the median corrective maintenance time depends on the distribution describing time to repair. Thus, for exponentially distributed repair time, the median corrective maintenance time is given by

$$T_{med} = (0.69) \text{ MTTR} \tag{3.3}$$

or

$$T_{med} = 0.69/\mu \tag{3.4}$$

where μ is the repair rate. For the exponential distribution, it is the reciprocal of the MTTR.

Similarly, for lognormally distributed repair time, the median corrective maintenance time is defined by

$$T_{med} = \text{MTTR}/\exp(\sigma^2/2) \tag{3.5}$$

where σ^2 is the variance around the mean value of the natural logarithm of repair times.

Maximum Corrective Maintenance Time

This measures the time required to complete all potential repair activities up to a given percentage, often the 90th or 95th percentiles.

For example, in the case of the 95th percentile, the maximum corrective maintenance time is the time within which 95% of all maintenance activities can be completed. In other words, no more than 5% of the maintenance activities will take longer than the maximum corrective maintenance time.

The calculation of maximum corrective maintenance time depends on the distribution used to describe the repair times. For three different repair time distributions, the maximum corrective maintenance times are as follows:

Lognormal Distribution

$$T_{mcm} = \text{antilog } (t_m + k\sigma) \tag{3.6}$$

where T_{mcm} is the maximum corrective maintenance time.

t_m is the mean of the logarithms of the repair times.

σ is the standard deviation of the logarithms of repair times.

k is equal to 1.28 or 1.65 for the 90th and 95th percentiles, respectively.

Exponential Distribution

For this distribution, T_{mcm} is approximately expressed by

$$T_{mcm} = 3 \text{ (MTTR)} \tag{3.7}$$

Normal Distribution

$$T_{mcm} = \text{MTTR} + k\sigma_n \tag{3.8}$$

where σ_n is the standard deviation of the normally distributed maintenance time.

Mean Maintenance Downtime

This is the total time needed either to restore equipment to a specified performance level or to maintain it at that level of performance. Thus it includes not only active corrective and preventive maintenance times but also administrative and logistic delay times. Administrative delay time is the equipment downtime created by some administrative constraint or priority. Logistic delay time is the time spent waiting for a required resource, such as a specific test, a spare part, or a facility.

Mean maintenance downtime is expressed by

$$T_{mmd} = T_{mam} + T_{ad} + T_{ld} \qquad (3.9)$$

where T_{mmd} is the mean maintenance downtime.

$\qquad T_{mam}$ is the mean active maintenance time, or mean time required to conduct corrective and preventive maintenance related tasks. (For formulas to compute this time, see References 2 and 4.)

$\qquad T_{ad}$ is the administrative delay time.

$\qquad T_{ld}$ is the logistic delay time.

MAINTAINABILITY FUNCTIONS

As they do in many other engineering disciplines, probability and statistics play an important role in maintainability. Various probability distributions may be used to present an item's repair time data. Once the repair time distributions are identified, the corresponding maintainability functions may be obtained. The maintainability functions are used to predict the probability that a repair, beginning at time $t = 0$, will be accomplished in a time t.

The maintainability function, m(t), for any distribution is expressed by

$$m(t) = \int_0^t f_r(t)dt \qquad (3.10)$$

where t is time.

$\qquad f_r(t)$ is the probability density function of the repair time.

The following are maintainability functions for the various probability distributions.

Exponential Distribution

This is a simple distribution to handle and is useful for presenting corrective maintenance times. The exponential distribution probability density function is defined by

$$f_r(t) = \left(\frac{1}{MTTR}\right) \exp\left(-\frac{t}{MTTR}\right) \tag{3.11}$$

where t is the variable repair time.
 MTTR is the mean time to repair.

By substituting Equation 3.11 into relationship 3.10, we get

$$m(t) = \int_0^t \left(\frac{1}{MTTR}\right) \exp\left(-\frac{1}{MTTR}\right) dt$$

$$= 1 - \exp\left(-\frac{t}{MTTR}\right) \tag{3.12}$$

Example 3-2

A mechanical system's mean time to repair (MTTR) is two hours. Calculate the probability that a repair will be completed in three hours, if the time to repair is exponentially distributed.
Substituting the specified data into Equation 3.12 yields

$$m(3) = 1 - \exp\left(-\frac{3}{2}\right)$$

$$= 0.7769$$

There is a likelihood of approximately 78% that the repair will be accomplished in three hours.

Lognormal Distribution

This is probably the most widely used probability distribution in maintainability work and is defined by

$$f_r(t) = \frac{1}{(t - \theta)\sigma\sqrt{2\pi}} \; \exp\left\{-\frac{1}{2}[\ln(t - \theta) - \beta]^2\right\} \qquad (3.13)$$

where θ is a constant denoting the shortest time below which no maintenance activity can be carried out.

σ is the standard deviation of the natural logarithm of the maintenance times around the mean value β.

β is the mean of the natural logarithm of the maintenance times.

The following relationship provides an estimate of the mean, β:

$$\beta = (\ln t_1 + \ln t_2 + \ln t_3 + \ldots + \ln t_m)/m \qquad (3.14)$$

where m is the number of maintenance activities performed.

t_i is the maintenance time i, for i = 1, 2, 3, . . . , m.

The standard deviation, σ, is given by the following relationship

$$\sigma = \left[\sum_{i=1}^{m}(\ln t_i - \beta)^2 / (m - 1)\right]^{1/2} \qquad (3.15)$$

The maintainability function, m(t), is expressed by

$$m(t) = \int_0^\infty t f_r(t) dt$$

$$= \frac{1}{\sigma\sqrt{2\pi}} \int_0^\infty \exp\left[-\frac{1}{2}\left(\frac{\ln t - \beta}{\sigma}\right)^2\right] dt \qquad (3.16)$$

Weibull Distribution

This is a distribution sometimes used to represent field maintenance times for complex electronic equipment. Under certain circumstances, the distribution of administrative delay times for field maintenance can be described by using the Weibull distribution [5].

The probability density function of the Weibull distribution is expressed by

$$f_r(t) = (b/\gamma^b)t^{b-1} \exp[-(t/\gamma)^b] \tag{3.17}$$

where γ is the scale parameter.
b is the shape parameter.

Inserting Equation 3.17 into Equation 3.10 yields

$$m(t) = \int_0^t (b/\gamma^b)t^{b-1} \exp[-(t/\gamma)^b]dt$$

$$= 1 - \exp[-(t/\gamma)^b] \tag{3.18}$$

At $b = 1$, Equation 3.18 reduces to

$$m(t) = 1 - \exp[-(t/\gamma)] \tag{3.19}$$

Equation 3.19 is for the exponential distribution and, for $\gamma = $ MTTR, it is identical to Equation 3.12. For $b = 2$, Equation 3.18 becomes the maintainability function for Rayleigh distribution. In this case, the equipment repair time is increasing at a linear rate.

The Weibull mean maintenance time, T_w, from Equation 3.19 and Reference 6 is

$$T_w = \int_0^\infty [1 - m(t)]dt$$

$$= \int_0^\infty (1 - \{1 - \exp[-(t/\gamma)^b]\})dt \tag{3.20}$$

$$= \gamma\Gamma\left(1 + \frac{1}{b}\right)$$

where $\Gamma\left(1 + \dfrac{1}{b}\right)$ is a form of the gamma function expressed as follows:

$$\Gamma(b) = \int_0^\infty t^{b-1} e^{-t} dt \tag{3.21}$$

Normal Distribution

This is another distribution that can be used to represent maintenance times. The probability density function of the normal distribution is defined by

$$f_r(t) = \frac{1}{\sigma\sqrt{2\pi}} \exp\left[-\frac{1}{2}\left(\frac{t-\mu}{\sigma}\right)^2\right] \tag{3.22}$$

where μ is the mean of maintenance times.

σ is the standard deviation of the variable maintenance time t around the mean value μ.

Inserting Equation 3.22 into Equation 3.10, we get the following maintainability function:

$$m(t) = \frac{1}{\sigma\sqrt{2\pi}} \int_{-\infty}^t \exp\left[-\frac{1}{2}\left(\frac{t-\mu}{\sigma}\right)^2\right] dt \tag{3.23}$$

The standard deviation, σ, is expressed by

$$\sigma = \left\{\sum_{i=1}^m (t_i - \mu)^2 /(m-1)\right\}^{1/2} \tag{3.24}$$

where m is the number of maintenance activities performed.

t_i is the maintenance time i, for i = 1, 2, 3, . . . , m.

The mean value, μ, of the maintenance times is given by

$$\mu = \sum_{i=1}^m t_i / m \tag{3.25}$$

Gamma Distribution

This is a two-parameter distribution used to represent various types of maintenance time data. The probability density function of the gamma distribution is expressed by

$$f_r(t) = \frac{c^b}{\Gamma(b)} t^{b-1} e^{-ct}$$ (3.26)

where b is the shape parameter.
           c is the scale parameter.

The gamma function, $\Gamma(b)$, is defined by

$$\Gamma(b) = \int_0^\infty x^{b-1} e^{-x} dx$$ (3.27)

Inserting Equation 3.26 into Equation 3.10 yields

$$m(t) = \frac{c^b}{\Gamma(b)} \int_0^t t^{b-1} e^{-ct} dt$$ (3.28)

For b = 1, Equation 3.28 becomes the maintainability function for the exponential distribution. The mean, M, of the gamma distributed maintenance times is

$$M = b/c$$ (3.29)

The standard deviation, σ, of the gamma distribution is given by

$$\sigma = (M/c)^{1/2}$$ (3.30)

Some of the special case values of Equation 3.27 are as follows:

- b = positive integer values $\Gamma(b) = (b - 1)!$
- b = 0, 1, 2, . . . $\Gamma(b + 1) = b!$
- b = 1 $\Gamma(1) = 1$
- b = 0.5 $\Gamma(0.5) = \sqrt{\pi}$

Erlangian Distribution

This is a special case of the gamma distribution in which the gamma distribution shape parameter takes positive integer values. In this case, Equation 3.27 yields

$$\Gamma(b) = (b - 1)! \tag{3.31}$$

The probability density function of the Erlangian distribution from Equation 3.26 is

$$f_r(t) = \frac{c^b}{(b - 1)!} t^{b-1} e^{-ct} \tag{3.32}$$

Substituting Equation 3.32 into Equation 3.10, we obtain

$$m(t) = 1 - \sum_{i=0}^{b-1} \{ e^{-ct} (ct)^i / i! \}$$

$$= \sum_{i=b}^{\infty} \{ e^{-ct} (ct)^i / i! \} \tag{3.33}$$

SYSTEM EFFECTIVENESS AND RELATED AVAILABILITY AND DEPENDABILITY MODELS

System effectiveness is the extent to which a system, product, or piece of equipment is capable of carrying out its assigned functions. The United States Army Material Command defines system effectiveness as the probability that a system can successfully satisfy an operational demand within a specified time period when used under designed conditions. There are various models and formulas that can represent aspects of effectiveness such as availability, dependability, and technical and performance parameters.

Availability Types

The types of availability include inherent availability, achieved availability, and operational availability. Inherent availability is a measure of the variables inherent in the design that affect availability. In the calculation of downtime it usually includes only active repair time; it does not include preventive maintenance time and administrative or logistic delay times. The inherent availability of a system or product is expressed by

$$IA = \frac{MTBF}{MTTR + MTBF} \tag{3.34}$$

where IA is the inherent availability.
    MTBF is the mean time between failures.
    MTTR is the mean time to repair.

Achieved availability is the probability that an item, when used under designed conditions in an ideal support environment, will perform satisfactorily. It includes both active repair time and preventive maintenance time but excludes administrative and logistic delay times. Achieved availability is expressed by

$$ACH = \frac{MTBM}{T_{mam} + MTBM} \tag{3.35}$$

where ACH is the achieved availability.
    MTBM is the mean time between corrective and preventive maintenance actions.

Operational availability is the probability that an item, when used under designed conditions in an actual operational environment, will perform satisfactorily. It includes active repair time, preventive maintenance time and administrative and logistic delay times. Operational availability is defined by

$$OAV = \frac{MTBM}{T_{mmd} + MTBM} \tag{3.36}$$

where OAV is the operational availability.

Dependability

Dependability is the measure of a system or product's condition during a mission, provided that it is operational and available at the beginning of the mission. Dependability can also be described as the probability that a system or product will accomplish its assigned mission, again provided that it was available for operation at the beginning of the mission. System reliability significantly impacts the dependability of an unmanned system/item.

Careful consideration given to maintainability and human factors during the design of manned systems and equipment can improve dependability. The dependability of a system or product is defined as [4]

$$D_s = OM(1 - OR) + OR \tag{3.37}$$

where D_s is the dependability.

OM is the operational maintainability and is expressed as probability that the system or product will be repaired or restored to a given operational state or retained in that state within a specified time period, when maintenance tasks are conducted by properly trained persons following the procedures prescribed.

OR is the operational reliability.

At OM = 0, Equation 3.37 reduces to complete operational reliability.

MATHEMATICAL MODELS

The numerous mathematical models directly or indirectly concerned with maintainability include models for determining required spare part quantity, estimating annual average maintenance labor hours, determining the probability of fault detection, and estimating mean time to fault detection.

Required Spare Part Quantity Estimation

In maintainability work, it is very important to determine the quantity of spare parts a system or product will require. This quantity depends upon many factors including the reliability of the system or product under consideration, the probability of having a spare part available when required, and the number of components that make up the system [4]. The following formula based on the Poisson distribution can be used to determine required spare part quantity:

$$SP = \sum_{i=0}^{m} [(-1)\ln(e^{-n\lambda t})]^i e^{-n\lambda t} / i! \qquad (3.38)$$

where SP is the probability of having a spare of a specific part available when required.

n is the number of spares of a specific type used.

t is time.

λ is the constant failure rate of a spare of a specific type.

m is the number of spare parts carried in stock.

Sometimes SP is called the safety factor, because it indicates the desired level of certainty that a spare part will be available when needed. As Equation 3.38 shows, the higher value of SP, the greater the quantity of spares required, and the higher purchasing and inventory costs will be.

Annual Mean Maintenance Labor-Hours Estimation

Annual mean maintenance labor-hours are the average total of preventive and corrective maintenance labor-hours expended over the course of a year. After determining MTTR and preventive maintenance times, the value of the annual mean maintenance labor-hours can be calculated using the following relationship [1]:

$$MLH_a = \frac{T_s(MTTR)(N_c)}{MTBF} + T_{pm}(N_p) \qquad (3.39)$$

where MLH_a is the mean maintenance labor-hours per year.

T_s is the number of operating hours per year.

MTBF is the mean time between failures.

MTTR is the mean time to repair.

N_c is the mean number of persons required to perform a corrective maintenance task.

T_{pm} is the annual mean preventive maintenance time.

N_p is the mean number of persons required to perform a preventive maintenance task.

Alternatively, Equation 3.39 can be expressed as follows:

$$MLH_a = (8760)\sum_{j=1}^{k} \frac{(N_{cj})(T_{dj})(MTTR_j)}{MTBF_j} + \sum_{i=1}^{n} (N_{pi})(f_i)T_{pi} \qquad (3.40)$$

where k is the total number of elements.

n is the total number of preventive maintenance tasks.

N_{cj} is the mean number of persons required to perform a corrective maintenance task to repair the jth element.

T_{dj} is the duty cycle time of jth element, in other words, the fraction of calendar time in which the jth element is operating.

$MTBF_j$ is the mean time between failures of the jth element, expressed in hours.

$MTTR_j$ is the mean time to repair of the jth element, expressed in hours.

N_{pi} is the mean number of persons required to perform the ith preventive maintenance task.

T_{pi} is the duration of the ith preventive maintenance task, expressed in hours.

f_i is the required frequency of ith preventive maintenance task.

Fault Detection Probability Estimation

Fault detection probability is an important piece of information as equipment availability depends upon fault detection accuracy. The probability of detecting faults correctly is expressed by [1]

$$PD_{cf} = \left(\sum_{j=1}^{k} \theta_j \lambda_{dj}\right) / \sum_{j=1}^{n} \lambda_j \qquad (3.41)$$

where n is the total number of items.

k is the total number of items possessing some degree of fault detectability.

λ_j is the failure rate of item j.

λ_{dj} is the detectable failure rate for item j.

θ_j is the portion of the failure rate that is detectable for item j.

PD_{cf} is the probability of detecting faults correctly.

Mean Time to Detect Estimation

When an item is dormant and cannot be monitored on a continuous basis, periodic testing will determine its suitability for an effective operation. Any failures that occur during dormancy are only detectable through periodic testing. The time to detect the failure of a dormant item may simply be described as the time between the failure and the test. The mean of such times is the mean time to detect and is defined by [1]:

$$\text{MTTD} = \frac{T}{(1 - e^{-\lambda_I T})} - \frac{1}{\lambda_I} \tag{3.42}$$

where MTTD is the mean time to detect.

T is the test period, expressed in hours.

λ_I is the item failure rate, given in failures per hour.

PROBLEMS

1. Compare corrective maintenance time and preventive maintenance time.
2. Three subsystems, i, j, and k, form an electronic system. The constant failure rates of these subsystems are $\lambda_i = 0.002$ failures per hour, $\lambda_j = 0.004$ failures per hour, and $\lambda_k = 0.006$ failures per hour. The corresponding estimated corrective maintenance times are $T_i = 2$ hours, $T_j = 3$ hours, and $T_k = 4$ hours, respectively. Estimate the mean time to repair (MTTR) for the overall system.
3. Discuss the following two items:
   • Maximum corrective maintenance time
   • Median corrective maintenance time
4. Define the maintainability function verbally and mathematically.

5. Obtain maintainability functions for the following distributions:
   - Exponential
   - Lognormal
6. After a detailed analysis of repair data associated with an engineering system, it was concluded that the system mean time to repair (MTTR) is 3.5 hours. Calculate the probability of completing a repair in 2.5 hours, if the times to repair are described by an exponential probability density function.
7. Prove that the mean, M, of the gamma distributed maintenance time is given by

   $$M = m/k$$

   where m is the shape parameter associated with gamma distribution.

   k is the scale parameter associated with the gamma distribution.
8. What is the difference between the following types of availability?
   - Inherent availability
   - Achieved availability
   - Operational availability
9. Describe the relationship between system effectiveness and dependability.
10. What are the important assumptions associated with the equation that determines the probability of having a spare of a specific item available when required?

REFERENCES

1. Grant-Ireson, W. and Coombs, C. F., editors. *Handbook of Reliability Engineering and Management.* McGraw-Hill Book Company, New York, 1988.
2. AMCP 706-133, *Maintainability Engineering Theory and Practice.* Department of Defense, Washington, D.C., 1976.
3. Blanchard, B. S. *Logistics Engineering and Management.* Prentice-Hall, Inc., Englewood Cliffs, New Jersey, 1981.
4. Blanchard, B. S., Verma, D. and Peterson, E. L. *Maintainability.* John Wiley and Sons, Inc., New York, 1995.
5. Von Alven, W. H., editor. *Reliability Engineering.* Prentice-Hall, Inc., Englewood Cliffs, New Jersey, 1964.
6. Dhillon, B. S. *Reliability Engineering in Systems Design and Operation.* Van Nostrand Reinhold Company, New York, 1983.

Maintainability Tools

INTRODUCTION

Over the years, reliability and maintainability professionals have developed various methods and techniques of analysis. Some of these approaches are only suitable for reliability, some only for maintainability, and some can be applied, with varying degrees of effectiveness, to either. Two methods applicable to both reliability and maintainability are failure mode and effects analysis (FMEA) and fault tree analysis (FTA). Both techniques were developed to handle reliability problems in defense and aerospace systems. They have since proven useful in many other industrial sectors and in addressing maintainability problems.

There are many other approaches to the various types of maintainability problems, among them cause and effect diagrams, total quality management (TQM), statistical control charts, and maintainability allocation determination.

FAILURE MODE, EFFECTS, AND CRITICALITY ANALYSIS

Failure mode and effects analysis (FMEA) is a structured qualitative analysis of a system, subsystem, component, or function that highlights potential failure modes, their causes, and the effects of a failure on system operation. When FMEA also evaluates the criticality of the failure, that is, the severity of the effect of the failure and the probability of its occurrence, the analysis is referred to as failure mode, effects, and criticality analysis (FMECA) and the failure modes are assigned priorities [1].

The FMEA technique was developed in the early 1950s to analyze flight control systems [2, 3]. During the course of the next decade, FMECA grew out of FMEA, and in the 1970s the United States Department of Defense developed a military standard [4] entitled *Procedures for Performing a Failure Mode, Effects, and Criticality Analysis.* The most important revision of the document occurred in 1984 [5]. See Reference 6 for a comprehensive list of publications on FMEA and FMECA.

There are three distinct types of FMECA: system level FMECA, design level FMECA, and process level FMECA. Of these three, the highest level of analysis is the system level FMECA, which usually consists of a collection of subsystem FMECAs. Performed in the initial design concept phase, the system level FMECA highlights potential system or subsystem failures so that they can be prevented. The design level FMECA helps identify and prevent failures stemming from the product design. It analyzes the design that has been developed and examines how failures of individual items would affect the system functioning or operation. The purpose of the process level FMECA is to analyze the process by which the product or system is to be built and assess how potential failures in the manufacturing or service process would affect the product/system functioning or operation. All three types of FMECA consist of the following basic steps [1]:

- Understanding system parts, operation, and mission
- Identifying the hierarchical, or identure, level at which the analysis is to be performed
- Defining each item expected to be analyzed—for example, component, module, or subsystem
- Establishing associated ground rules and assumptions—for example, system mission and operational phases
- Identifying possible failure modes for each item
- Determining the effect of each item's failure for every possible failure mode
- Determining the effect of group failures—failures of more than one item—on system operation and mission
- Identifying methods, procedures, or approaches for detecting potential failures
- Determining any provisions or design changes that would prevent failures or mitigate their effects

FMECA Users and Information Needs

The FMECA must satisfy the needs of groups, among them design, reliability, maintainability, manufacturing, systems, testing, quality assurance, system safety, and integrated logistics support staff; prime contractors; the company's internal regulatory agency or customer representatives; and government agencies.

The information required to perform an FMECA includes, in the case of design-related information, equipment and part drawings; functional block diagrams; design descriptions and design-change history; system schematics; narrative descriptions; operating specifications and limits; configuration management data; relevant military, commercial, company, and/or customer specifications; interface specifications; guidelines for the design under consideration; effects that environmental factors such as temperature, humidity, vibration, dust, moisture, and radiation have on part and equipment reliability; part failure rates; field service data; and reliability data, including historical data on failures and cause and effect analyses of previous failures. Information related to an FMECA and sources for obtaining it are:

- Item identification numbers, available from the parts list for the system or product
- Item nomenclature/functional specifications, available from the design engineer or from the parts list
- System or product function, available in the customer requirements or from the design engineer
- Provisions or design changes to prevent or compensate for failures, available from the design engineer
- Mission phase/operational mode, available from the design engineer
- Failure effects, available from the safety engineer, design engineer, and reliability engineer
- Failure modes, causes, and rates, available from the factory database and the field experience database
- Failure probability/severity classification, available from the safety engineer
- Failure detection method(s) available from the maintainability engineer and the design engineer

Criticality Assessment

This assessment ranks potential failures identified during the system analysis based on the severity of their effects and the likelihood of their occurrence. The two methods most often used for making a criticality assessment are risk priority number (RPN) method and military standard method.

Risk Priority Number Method

This technique, commonly used in the automotive industry, bases the risk priority number for an item failure mode on three factors: probability of occurrence, the severity of the failure's effects, and probability of failure detection. The probability of occurrence is the likelihood of failure, or relative number of failures, expected during the item's useful life. Table 4.1 describes the rankings of probability of occurrence [7]. The severity of effect of an item's failure is the consequences it will have for the next highest level of the system, the system as a whole, and/or the user. Table 4.2 describes the rankings of severity of effect [7]. The probability of failure detection is an assessment of the proposed design verification program's ability to

Table 4.1
Rankings of Probability of Occurrence and Associated Descriptions

Description of Ranking	Probability of Occurrence	Rank
Very high (the failure is very likely to occur	1 in 2	10
Very high	1 in 8	9
High (the failure will occur often)	1 in 20	8
High	1 in 40	7
Moderate (the failure will occur occasionally)	1 in 80	6
Moderate	1 in 400	5
Moderate	1 in 1,000	4
Low (the failure will rarely occur)	1 in 4,000	3
Low	1 in 20,000	2
Remote (the failure is unlikely to occur)	<1 in 10^6	1

Table 4.2
Rankings of Severity of Failure Effect and Associated Descriptions

Level of Severity	Rank
Very high (the failure will affect safe product operation)	9, 10
High (there will be a high degree of customer dissatisfaction because of the failure)	7, 8
Moderate (the failure will generate some customer dissatisfaction)	4, 5, 6
Low (the failure will only cause minor customer annoyance)	2, 3
Minor (customer may not even become aware of the failure)	1

Table 4.3
Rankings of Likelihood of Detection and Associated Descriptions

Likelihood of Detection	Rank
Non-detection inevitable (potential design problems cannot be detected by the program)	10
Very low (program probably will not be able to detect a potential design problem)	9
Low (program is unlikely to detect a potential design problem)	7, 8
Moderate (program may detect a potential design problem)	5, 6
High (there is a good chance that the program will detect a potential design problem)	3, 4
Very high (it is almost certain that the program will detect a potential design problem)	1, 2

detect a potential problem before the item involved goes into production. Table 4.3 describes the rankings of probability of detection [7].

The risk priority number is expressed by

$$RPN = (OR)\ (SR)\ (DR) \tag{4.1}$$

where OR is the ranking of probability of occurrence.
    SR is the ranking of severity of effects.
    DR is the ranking of probability of detection.

Failure modes with a high RPN are more critical and given a higher priority than ones with a lower RPN. When the scales used range from 1 to 10, the value of an RPN will be between 1 and 1,000. The scales and categories used may, of course, vary from one organization to another.

Military Standard Method

The Department of Defense, in *Procedures for Performing a Failure Mode, Effects, and Criticality Analysis* [5] set forward a technique for ranking potential failure modes that is often used in the defense, aerospace, and nuclear power generation industries. The military standard method consists of distinct qualitative and quantitative approaches. The qualitative approach, used when failure rate data are not available, groups occurrence probabilities for individual item failures together into levels that establish qualitative failure probabilities.

Table 4.4 presents the set of levels and associated guidelines used in the military standard method. After the failure-mode probability level is determined, the probability level and severity classification of the failure mode are plotted on a criticality matrix, as shown in Figure 4-1. Table 4.5 presents the failure mode severity classifications.

Table 4.4
Qualitative Ranking of Failure Probabilities

Level of Probability of Occurrence	Short Description of the Rank Level	Detailed Description of the Rank Level
V	Extremely unlikely	The probability of a failure during the item's functional period is virtually negligible.
IV	Remote	The probability of a failure during the item's functional period is remote.
III	Low to moderate	The probability of a failure during the item's functional period is low to moderate.
II	Moderate	The probability of a failure during the item's functional period is moderate.
I	High	The probability of a failure during the item's functional period is high.

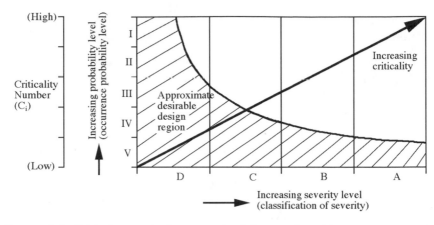

Figure 4-1. Criticality matrix for comparing failure modes with respect to severity.

Table 4.5
Classification of Failure-mode Severity

Severity Classification	Short Description of the Classification	Detailed Description of the Classification
D	Minor	The failure will lead to unscheduled maintenance or repair but will not be serious enough to result in injury, property damage, or system damage.
C	Marginal	The failure will lead to delay or loss of availability or mission degradation and may also cause minor injury, minor property damage, or minor system damage.
B	Critical	The failure will lead to mission loss and may also cause severe injury, major property damage, or major system damage.
A	Catastrophic	The failure may result in death or system loss.

The criticality matrix presented in Figure 4-1 provides a mechanism for comparing the probability and severity of failure modes. The criticality matrix represents the combined factors of the severity of the potential failure's effects and the probability that the failure will occur. This matrix can help set priorities for addressing potential failures and developing appropriate corrective measures. The area of the matrix labeled "approximate desirable design region" indicates a low probability of failures with class A and B severity effects and anywhere from a low to high probability of class C and D failures that can be tolerated. Nonetheless, every possible step should be taken to eliminate class A and B failure modes, or at least to reduce their probability of occurrence, by making appropriate design changes.

The quantitative approach, used when failure rate data are available, defines the failure-mode criticality number, N_{cf}, by

$$N_{cf} = \lambda_p T \theta n \qquad (4.2)$$

where λ_p is the constant failure rate of the item.

$\quad$ T is the item operating time.

$\quad \theta$ is the conditional probability that the effect of the failure will match the identified severity classification. Table 4.6 presents quantified values for θ.

$\quad$ n is the failure mode apportionment ratio, or the probability that the item will fail in the specific failure mode under consideration. In other words, it is the fraction of the item failure rate that can be apportioned to the failure mode of interest. Furthermore, when all failure modes of an item are specified, the sum or addition of the apportionments is equal to unity. Table 4.7 presents examples of failure mode apportionment ratios.

The item criticality number, C_i, is calculated for each severity class. It is the total of the critical numbers associated with each of the item's failure modes that fall into the severity class under consideration:

$$C_i = \sum_{i=1}^{k} (N_{cf})_i = \sum_{i=1}^{k} (\lambda_p T \theta n)_i \qquad (4.3)$$

where k is the number of item failure modes that fall into the severity classification under consideration.

Table 4.6
Failure Effect Probability Values

Failure Effect Description	Value for θ (probability)
No effect	0
Possible loss	Between 0 and 0.10
Probable loss	Between 0.10 and 1.00
Actual loss	1

Table 4.7
Examples of Part Failure Mode Apportionments

Item Description	Item Failure Mode	Apportionment Value (or probability value for n)
Hydraulic valve	a) Stuck closed	0.12
	b) Stuck open	0.11
	c) Leaking	0.77
Variable resistor	a) Open	0.53
	b) Short	0.07
	c) Erratic output	0.40
Relief valve	a) Prematurely open	0.77
	b) Leaking	0.23
Fixed resistor	a) Short	0.05
	b) Open	0.84
	c) Parameter change	0.11

When an item failure mode results in multiple severity-class effects, each with its own occurrence probability, only the most critical should be used in the computation C_i [8]. Otherwise, the result may be mistakenly low values of C_i for the less critical severity classes. Therefore, θ values should be calculated for all severity classes associated with a failure mode, including those associated with class B, C, and D failures.

FMECA Benefits

Some of the advantages of performing an FMECA are that it [9]:

• Proves useful for making design comparisons
• Generates input data for use in test planning

- Serves as a visibility tool for managers
- Provides a systematic approach to classifying hardware failures
- Identifies all possible failure modes and their effects on mission, personnel, and system
- Generates useful data for use in system safety and maintainability analyses
- Helps improve communication among design interface personnel
- Effectively analyzes small, large, and complex systems
- Is easy to understand
- Starts from the level of greatest detail and works upward

FAULT TREE ANALYSIS

This powerful reliability analysis tool can be used for various maintainability-related problems. Fault tree analysis (FTA) defines an undesirable state of the system or product and then analyzes the system or product, in terms of its operation and environment, to determine all possible ways in which the undesirable event can occur. An FTA is a useful tool to identify all possible failure causes at all possible levels associated with a system and to identify the relationship between causes. It can thus improve the design of any specified system, product, or process. An FTA normally takes place during the early design phase and then is progressively refined and updated as the design develops.

Bell Laboratories developed the FTA technique in the early 1960s to evaluate the reliability and safety of the Minuteman Launch Control System. Reference 10 describes the method in detail and Reference 11 gives a comprehensive list of sources on the subject.

Fault Tree Logic and Event Symbols

There are many logic and event symbols used to construct fault trees [10]. Figure 4-2 presents the two most commonly used logic symbols: the OR gate and AND gate.

The OR gate symbol signifies that an output fault event occurs if one or more of the m input fault events occur. The AND gate symbol denotes that the output fault event only occurs if all of the m input fault events occur.

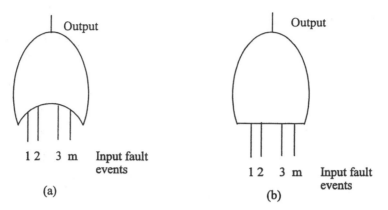

Figure 4-2 Two commonly used fault tree logic symbols: (a) OR gate, (b) AND gate.

Figure 4-3 shows two frequently used fault event symbols, the circle and the rectangle. The circle denotes the failure of an elementary component or a basic fault event that need not be traced back any further. The fault parameters such as the failure rate, the probability of occurrence, and the repair rate can be obtained from the field failure data or other similar sources. The rectangle denotes a fault event that results from a combination of preceding fault events.

Objectives and Prerequisites

Fault tree analysis of a system can be used to: identify critical areas and cost-effective improvements; provide input to testing, maintenance, and operational procedures and policies; confirm the ability of the system to fulfill its imposed safety requirements; meet jurisdictional requirements; provide input for cost-benefit analysis of trade-offs; evaluate performance of systems or products for bid-evaluation purposes; and highlight requirements or targets for systems.

The prerequisites for a fault tree analysis include clearly defined analysis scope and objectives, clear identification of assumptions, well-defined level of analysis resolution, thorough understanding of the system's design and its operation and maintenance aspects, well-defined physical bounds and interfaces for the system, a comprehensive review of system operational experience, and a clear definition of what constitutes system failure or undesirable events.

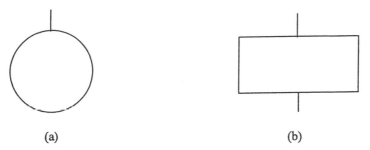

(a)   (b)

Figure 4-3 Two commonly used fault event symbols: (a) circle; (b) rectangle.

Fault Tree Analysis Steps

The basic steps involved in performing fault tree analysis are as follows [12]:

- Define the system, the assumptions involved in the analysis, and the events or states that would constitute failure.
- If simplification of the scope of analysis is desirable, establish a system block diagram indicating inputs, outputs, and interfaces.
- Establish the top-level fault event.
- Use fault tree logic and fault event symbols and apply deductive reasoning to identify what could cause the top-level fault event to occur.
- Continue developing the logic tree by identifying causes for intermediate fault events, that is, the fault events that can cause the top-level fault event to occur.
- Develop the fault tree to the desired lowest level, that of the most basic fault events.
- Analyze the completed fault tree qualitatively and quantitatively.
- Identify appropriate corrective measures.
- Document the analysis and take appropriate measures to rectify problem areas.

Example 4-1

Assume that a workshop repairs failed electric motors. Develop a fault tree for the following undesired event—an electric motor will not be repaired by a given point in time—using the fault tree symbols described earlier. Consider only these four factors:

- Motor is too damaged to repair
- Skilled manpower is unavailable
- Spare parts are unavailable
- Repair tools or facilities are unavailable

Figure 4-4 shows the fault tree for this example. In this figure, T identifies the top-level fault event; I identifies the intermediate fault event; and B_i identifies the basic or primary fault event, for i = 1, 2, 3, 4, 5. The top-level event, T, will occur if any of the fault events that appear within a circle—B_1, B_2, B_3, B_4, or B_5— occurs.

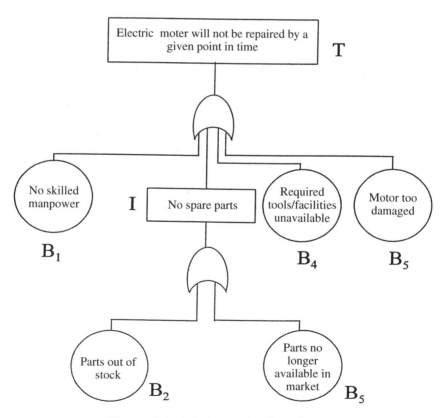

Figure 4-4. A fault tree for Example 4-1.

Fault Tree Probability Evaluation

When the occurrence probability of basic or other fault events is available, the probability of occurrence of the top-level event can be obtained for fault trees containing AND and OR gates. The probability evaluation for both these gates appears in the following [10].

OR Gate

Figure 4-5 shows an m input fault events OR gate. The probability of occurrence of the OR gate output fault event X is given by

$$P(X) = 1 - \prod_{i=1}^{m} [1 - P(x_i)] \tag{4.4}$$

where P(X) is the probability of occurrence of the OR gate output fault event X.

$P(x_i)$ is the probability of occurrence of the x_i independent input fault event, for i = 1, 2, 3, . . . , m.

m is the number of input fault events.

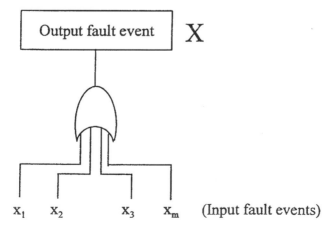

Figure 4-5. An m input fault events OR gate.

For m = 2, Equation 4.4 reduces to

$$P(X) = 1 - \prod_{i=1}^{2} [1 - P(x_i)]$$

$$= 1 - [1 - P(x_1)][1 - P(x_2)] \tag{4.5}$$

$$= P(x_1) + P(x_2) - P(x_1)P(x_2)$$

If the probability of occurrence of each input fault event is less than 0.1, Equation 4.4 may be approximated, as can be seen from Equation 4-5, by the following expression:

$$P(X) \cong \sum_{i=1}^{m} P(x_i) \tag{4.6}$$

AND Gate

Figure 4-6 shows an m input fault events AND gate. The probability of occurrence of the AND gate fault event Y is given by

$$P(Y) = \sum_{i=1}^{m} P(y_i) \tag{4.7}$$

where P(Y) is the probability of occurrence of the AND gate output fault event Y.

P(y_i) is the probability of occurrence of the y_i independent input fault event, for i = 1, 2, 3, . . . , m.

Example 4-2

Assume that the probability of occurrence of Figure 4-4 basic fault events B_1, B_2, B_3, B_4, and B_5 are as shown in Figure 4-7. Using Equations 4.4 and 4.7, calculate the probability of occurrence of the top-level fault event T: electric motor will not be repaired in a given point in time.

Substituting the specified values for fault events B_2 and B_3 into Equation 4.4 yields the probability of occurrence of the intermediate fault event I:

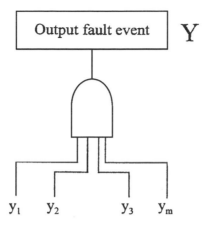

Figure 4-6. An m input fault events AND gate.

$P(I) = 0.02 + 0.03 - (0.02)\ (0.03)$

$P(I) = 0.0494$

Inserting into Equation 4.4 this result, and the other specified data for basic fault events B_1, B_4, and B_5 given in Figure 4-7, the top-level fault event T occurrence probability is given by

$P(T) = 1 - (1 - 0.0494)\ (1 - 0.01)\ (1 - 0.04)\ (1 - 0.05)$

$= 1 - (0.9506)\ (0.99)\ (0.96)\ (0.905)$

$= 0.1417$

The occurrence probability of the top-level event, the electric motor will not be repaired by a given point in time, is 0.1417.

FTA Advantages and Disadvantages

As is the case with any technique, fault tree analysis has certain advantages and disadvantages. Some of the advantages, as mentioned, are that it provides insight into system behavior; can handle complex systems more easily than some other techniques; provides a visibility tool that management, designers, and users can use to justify design changes and trade-off studies; and permits either qualitative or quantitative

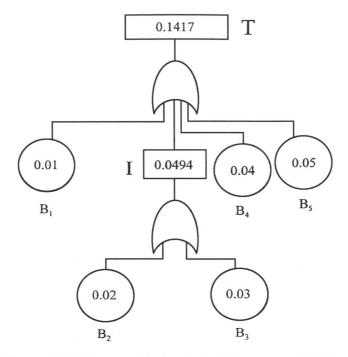

Figure 4-7. Fault tree with given basic fault event probabilities.

analysis. But it can also be a costly, time-consuming method of analysis, which has difficulty in handling states of partial failure and whose results can be difficult to check.

CAUSE AND EFFECT DIAGRAM

This technique is also known as an Ishikawa diagram, after its originator, Professor K. Ishikawa of Japan, or as a "fishbone" diagram. The latter term comes from the fact that the diagram resembles the skeleton of a fish. The cause and effect diagram is a deductive analytical approach useful in maintainability work. It makes use of a graphic "fishbone" to depict the cause and effect relationships between an undesired event and its associated contributing causes. In other words, it simply displays the factors that cause an effect such as a given problem: the effect represented can also be something positive, such as a goal to be attained.

Figure 4-8 shows a typical cause and effect, or fishbone, diagram.
The box, or the "fish head," on the right side represents the effect
(the problem or goal). The dotted box on the left side contains "fish
bones" that can be any set of factors considered to be important
causes. For example, in the cause of a control valve failure, the effect
could be "control valve failed to operate," and the main cause factors
could be manpower, materials, machinery, methods, and environment.

Examples of subcause factors for each of these main cause factors
might be [12]:

- **Manpower:** assembly error, insufficient manpower
- **Materials:** defective value, wrong material for valve parts
- **Machinery:** loss of power, leak in piping system
- **Methods:** faulty circuit, no input signal
- **Environment:** contamination, too cold or too hot

The subcause factors may have any number of subfactors of their
own. The basic steps in developing a cause and effect diagram are [13]:

- Establishing a problem statement or identify the effect to
  be investigated.
- Conducting brainstorming session(s) to identify possible causes.
- Grouping important causes into categories and stratify them.
- Developing the diagram.
- Refining the categories by asking questions such as "Why does
  this condition exist?" and "What causes this?"

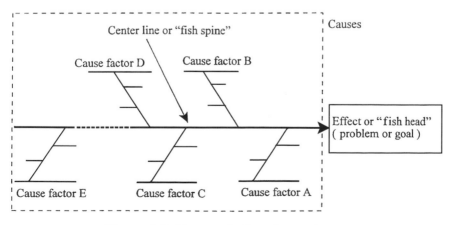

Figure 4-8. Cause and effect diagram layout.

A cause and effect diagram, like a fault tree analysis, seeks to identify the causes of problems, but a fault tree analysis examines causes in a more highly structured manner.

TOTAL QUALITY MANAGEMENT

Total quality management (TQM) is a philosophy of pursuing continuous improvement in every process, through the integrated or team efforts of all individuals associated with an organization. It has proven useful to organizations seeking to improve the maintainability of their products.

There is a commonly held belief that the TQM concept originated in Japan. While it was tested and enriched in that country, TQM can be traced to the work of quality experts such as W. E. Deming, J. Juran, and A. V. Feigenbaum in the late 1940s [13]. The Japanese Union of Scientists and Engineers (JUSE) asked General Douglas MacArthur to bring W. E. Deming to lecture in Japan. In July 1950, Deming delivered his first lecture on the elementary principles of statistical control of quality to an audience of Japanese engineers and scientists [14]. The next year, JUSE established the Deming prize, to be awarded to the company that had most effectively implemented quality control policies and measures [15].

In 1985, an American behavioral scientist, Nancy Warren, coined the term Total Quality Management [16]. Two years later, the United States government established the Malcolm Baldrige Award after witnessing the success of the Deming prize in Japan. The first Malcolm Baldrige Award went to the cellular telephone division of Motorola in 1988, for its reduction of defects from $1,000/10^6$ to $100/10^6$ between 1985 and 1988 [17, 18]. Over the years many people have contributed to the development of TQM, and a large number of publications are available on the subject.

Traditional Quality Assurance Programs (TQAP) versus TQM

In order to understand the TQM concept, it is important to distinguish between it and traditional quality assurance programs (TQAP).

A comparison between the two, on the basis of some key factors, follows [15, 18]:

- **Definition of program**
  - **TQAP:** product-driven
  - **TQM:** customer-driven

- **Objective of program**
  - **TQAP:** to find faults
  - **TQM:** to prevent faults

- **Definition of quality**
  - **TQAP:** products satisfy specifications
  - **TQM:** products are suitable for consumer use and/or application

- **Cost**
  - **TQAP:** improved quality results in greater cost
  - **TQM:** improved quality lower cost and increases productivity

- **Approach to customer requirements**
  - **TQAP:** ambiguous comprehension of customer requirements
  - **TQM:** well-defined mechanism to understand and satisfy customer requirements

- **Responsibility for quality**
  - **TQAP:** responsibility belongs to the quality control department or inspection center
  - **TQM:** every individual in the organization shares responsibility

- **Decision making**
  - **TQAP:** top-down approach
  - **TQM:** team approach, with each employee part of a team

TQM Principles and Elements

Two fundamental principles of TQM are customer satisfaction, whether the customer is inside or outside the organization, and continuous improvement. The use of this "market-in" concept allows a healthy customer orientation that recognizes that every work process is composed of stages [19]. Consequently, at each stage TQM seeks customer input to determine changes that would help better fulfill customers' needs.

The other principle, continuous improvement, requires that management continuously improve, and strive to make important breakthroughs in, the techniques, procedures, and processes used. The important elements of TQM fall into seven distinct categories, as shown in Figure 4-9 [19].

Management commitment and leadership is critical to the success of a TQM effort. Senior management must thoroughly understand the TQM concept to play a leadership role. Only then can management establish effective new goals and directions for the organization and play the leadership role in achieving those goals and directions.

A firm's ability to produce a quality product depends in part on the relationship among all parties involved. Thus supplier participation is an important element of TQM. To achieve supplier responsiveness, partnership and mutual trust are key. Some companies also require their suppliers to have formal TQM programs.

The team approach involves all parties, including vendors, customers, and subcontractors, in the TQM effort. The quality team can range in size from 3 to 15 members who usually serve on a voluntary basis. The team members possess skills in areas such as cost-benefit analysis, statistics, brainstorming, planning and controlling projects, and creating flow charts. The team leader, who usually belongs to management, chairs the team meetings.

As the demand for higher quality increases, customer involvement becomes increasingly important. A representative of customer service should therefore be part of the quality team, and customer interests,

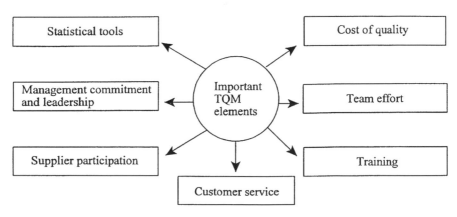

Figure 4-9. Important elements of TQM.

like other such interests, should be reflected in the goals, plans, and controls that the quality team develops.

A basic quality measurement tool is the cost of quality. This tool can be applied in monitoring the effectiveness of the TQM process, selecting quality improvement projects, and justifying the cost to doubters. It is an effective tool for raising awareness of quality issues and for communicating to management the benefits of the TQM concept in terms of dollars.

A Japanese axiom states that "quality starts with training and ends with training" [20]. Under TQM, each individual employed by the company is responsible for quality. The training effort must therefore be targeted at every level of the company, and specifically tailored for the needs of management, engineers, support personnel, field labor, technicians, and other groups.

Statistical tools useful in solving TQM-related problems, include control charts, Pareto diagrams, cause and effect diagrams, scatter diagrams, flow charts, and graphs and histograms. Some of their applications are [19, 21]:

- Identifying and classifying the causes of quality problems
- Verifying, repeating, and reproducing measurements based on data
- Making decisions on facts based on data rather than the opinions of individuals
- Communicating information in a language that can easily be understood by all quality team members

Deming's TQM Philosophy and the "Deadly Diseases" of American Management

W. E. Deming's TQM philosophy is based on fourteen points [22]: create a constancy of purpose towards improvement; adopt the philosophy "quality first"; drive out fear; institute leadership; end dependence on mass inspection and instead prevent defects; take action toward change; eliminate numerical quotas; end awards on cost savings alone; remove barriers to pride of workmanship; institute training in job skills; improve the system constantly and always; institute a vigorous program of education and retraining; break down barriers between staff; and eliminate slogans, exhortations, and targets.

The seven "deadly diseases" of American management, as defined by Deming, are lack of constancy of purpose, excessive liability costs (the U.S. is the world leader in lawsuits), excessive medical costs (General Motor's highest-paid supplier is Blue Cross), overemphasis on short-term profits, mobility of managers (which works against understanding and long-term efforts), performance reviews (which destroy teamwork and build fear), and management based on visible figures alone (the effects of a dissatisfied customer cannot be measured) [22].

TQM Tools

There are a large number of analytical approaches, both basic and advanced, that can be used in practicing TQM. The following seven analytical approaches belong to the basic category [22–23]:

- **Cause and effect diagram**—Simply displays the factors that cause a given effect such as a problem to be addressed or a goal to be attained. This approach was described earlier in the chapter.
- **Pareto chart**—Shows graphically the relative magnitude of output from different factors.
- **Checklist**—A means of collecting data, especially of recording events that occur against a list of possible events.
- **Control chart**—Highlights statistically significant changes that may happen in a process. In other words, a control chart is a graphic presentation of data collected over a time period that show upper and lower limits for the process to be controlled.
- **Flow chart**—A diagram showing the inputs and outputs of all operations in a process.
- **Scattergram**—A test for correlation between two factors. For each point, data is collected, and then the value of one factor is plotted vertically and the other horizontally.
- **Histogram**—Employed to show the distribution of outcomes of a repetitive event.

Analytical approaches belonging to the advanced category include:

- **Quality function deployment (QFD)**—Based on a matrix that compares "what" to "how." More specifically, QFD is a formal process used for translating customer requirements into appropriate technical requirements.

- **Quality loss function**—A concept developed by Genichi Taguchi, a Japanese statistician, and is based on the premise that the cost of quality increases not only when the finished product falls outside certain specifications but also when it deviates from a set target value within the specifications.

TQM Implementation and Related Pitfalls

The implementation of TQM involves five basic steps: creating a vision, planning an action, creating a structure (for example, instituting appropriate training, involving employees, eliminating roadblocks, and creating cross-functional teams), measuring progress, and updating the plans and vision as appropriate.

Some difficulties many firms have had in implementing TQM include: failure of top management to devote sufficient time to the effort, management insistence concerning implementing processes in a way employees find unacceptable, unsatisfactory allocation of resources for developing and training manpower, and failure of senior management to delegate decision-making authority to lower organizational levels [13].

MAINTAINABILITY ALLOCATION

Maintainability allocation may be described as analysis of equipment or system design, with the goal of allocating specified quantitative maintainability requirements down to the level of subsystems and/or components. These allocations relate to corrective maintenance characteristics and throughout the product development program help determine how well the design meets the overall maintainability requirements. They are a design and management tool that the customer, prime contractor, and/or subcontractor uses to [24]:

- Provide maintainability and design professionals with a document for monitoring and evaluating subcontractors' compliance with specified maintainability goals.
- Establish "not to exceed" maintainability targets for the product's or system's lower-level elements.
- Highlight aspects of the system that will require high levels of maintenance work, so that these problem areas can be analyzed

and the design changes that will have the greatest maintainability impact can be made.

Maintainability allocations for a complex system begin during the early design phase and are revised as appropriate throughout the development cycle. The maintainability allocation process begins with an examination of the system's design, which is then broken down into subsystem, assembly, and component levels. The specified maintainability values are allocated first from system to subsystem levels and then to progressively lower levels as the design develops.

Parameters such as mean time to repair (MTTR), maximum corrective maintenance time, and maintenance manhours per operating hour are often specified at the system level. Among the various approaches to maintainability allocation is the following:

The maintainability allocation method makes use of various weighting factors, which are related to environment, accessibility, handling, packaging, fault isolation techniques, complexity, and other issues, in allocating specified mean active corrective maintenance times for the system, subsystem, and component levels. The items with high failure rates are allocated the least time, and those with the lowest failure rates are allocated the most time. This is based on the premise that the more often an item is maintained or repaired, the less time it takes to do so. A weighting factor, θ_j, for a particular item is expressed by [25, 26]

$$\theta_j = \sum_{i=1}^{k} \theta_i / k \tag{4.8}$$

where k is the number of weighting factors associated with an item.
θ_i is the value of the weighting factor i, for i = 1, 2, 3, . . . k.

Assuming that each sub-item for which there will be an allocation has a weighting factor, the weighting factor for the item as a whole is given by

$$\theta_t = \sum_{j=1}^{n} \lambda_j \theta_j / \lambda \tag{4.9}$$

where θ_j is the sum of weighting factors for the item.
n is the number of sub-items.

λ_j is the failure of sub-item j.

$$\lambda = \sum_{j=1}^{n} \lambda_j$$

Solving Equation 4.9 for λ, we get

$$\lambda = \sum_{j-1}^{n} \lambda_j \theta_j / \theta_t \qquad (4.10)$$

Now consider that the mean active corrective maintenance time of an item is expressed by

$$\overline{T}_{ac} = \sum_{j=1}^{n} \lambda_j \overline{T}_{acj} / \lambda \qquad (4.11)$$

where $\overline{T}_{acj}$ is the mean active corrective maintenance time of sub-item j.

Rearranging Equation 4.11, we get

$$\lambda = \sum_{j=1}^{n} \lambda_i \overline{T}_{acj} / \overline{T}_{ac} \qquad (4.12)$$

Equating Equations 4.10 and 4.12 results in

$$\left(\sum_{j=1}^{n} \lambda_j \theta_j / \theta_t \right) = \left(\sum_{j=1}^{n} \lambda_i \overline{T}_{acj} / \overline{T}_{ac} \right) \qquad (4.13)$$

Rearranging Equation 4.13 yields

$$\lambda_1 \left(\frac{\theta_1}{\theta_t} - \frac{\overline{T}_{ac1}}{\overline{T}_{ac}} \right) + \star + \lambda_n \left(\frac{\theta_n}{\theta_t} - \frac{\overline{T}_{acn}}{\overline{T}_{ac}} \right) = 0 \qquad (4.14)$$

Thus,

$$\frac{\theta_j}{\theta_t} - \frac{\overline{T}_{acj}}{\overline{T}_{ac}} = 0$$

Therefore,

$$\overline{T}_{acj} = \theta_j \overline{T}_{ac} / \theta_t \tag{4.15}$$

Tables 4.8 to 4.15 present examples of factors for accessibility, complexity, environment, packaging, fault isolation, and handling.

Table 4.8
Selected Accessibility Factor Values

Factor Type	Factor Value (θ_1)	Considerations
Difficult	4	Screw cover
Simple	2	Quick release cover fasteners
Direct	1	No cover

Table 4.9
Selected Complexity Factor Values for Units

Unit Type	Factor Value (θ_2)	Considerations
Simple Power Supply	2	—
Complex power supply	3	—
Digital computer	2	Inclusive of power supply
Keyboard	2.5	Inclusive of code/decode and lamp drivers

Table 4.10
Selected Complexity Factor Values for Parts

Part Type	Factor Value (θ_3)	Considerations
Switch	2	—
Variable resistor	2	—
Transformer	3	—
Relay	3	Mechanical
Lamp	1	—

Table 4.11
Selected Complexity Factor Values for Assemblies

Assembly Type	Factor Value (θ_4)	Considerations
Digital circuit card assembly	1	Small
Low level analog circuit assembly	2	Large
Control	3	Complete
Simple power supply	2	Complete

Table 4.12
Selected Environment Factor Values

Factor Type	Factor Value (θ_5)	Considerations
Mild	1	Normal indoor temperature and relative humidity
Mild	2	Outdoor temperature: : 50–90°F Relative humidity: <90% Zero wind and precipitation
Severe	3	Indoor temperature: between 32°F and 100°F
Severe	6	Moderate outdoor tempeature but rain or snow and heavy wind

Table 4.13
Selected Package Factor Values

Factor Type	Factor Value (θ_6)	Considerations
Quick disconnect circuit card assembly	1	No fasteners
Circuit card assembly with solder connections	6	With screw fasteners
Pigtail connector circuit card assembly	2	With screw fasteners

Table 4.14
Selected Fault Isolation Technique Factor Values

Factor Type	Factor Value (θ_7)	Considerations
Manual	0.5	Using portable test equipment at circuit test points to make measurements manually
Automatic	1	Built-in computerized test
Semi-automatic	3	Built-in test circuits, controlled manually

Table 4.15
Selected Handling Factor Values

Factor Type	Factor Value (θ_8)	Considerations
Difficult (awkward)	3	Two persons required
Simple (light weight)	1	Only one person required

Example 4-3

Assume that a system is made up of units A, B, and C and that its specified mean active corrective maintenance time, $\overline{T}_{ac}$, is 0.5 hour. The estimated failure rates of units A, B, and C, along with their applicable values for θ_1, θ_2, θ_5, θ_7, and θ_8, are specified in Table 4.16. Allocate system $\overline{T}_{ac}$ among its units by using the given information.

Using the data given in Table 4.16 and Equation 4.8, we obtain the values for units A, B, and C as presented in Table 4.17.

Using the data given in Table 4.16, we get the following failure rate for the entire system:

$$\lambda = 50 + 25 + 10$$

$$= 85 \text{ failures}/10^6$$

where λ is the system failure rate.

Substituting this value, and the one obtained in Table 4.7, into Equation 4.9 yields

$$\theta_t = \sum \lambda_j \theta_j / \lambda$$

$$= \frac{167}{85} = 1.9647$$

Inserting the given and calculated values into Equation 4.15, we get

$$T_{acj} = \theta_j (0.5) / (1.9647)$$

$$= (0.2545) \theta_j \tag{4.16}$$

When the calculated values for θ_j given in Table 4.17 are substituted into Equation 4.16, we get the allocated mean active corrective maintenance time for units A, B, and C, presented in Table 4.18.

Table 4.16
Failure Rate and Factor Values for Example 4-3

Unit	Failure Rate, λ_j (failures/10^6)	θ_1	θ_2	θ_5	θ_7	θ_8
A	50	2	2	1	0.5	3
B	25	4	2	3	1	3
C	10	1	2.5	1	3	1

Table 4.17
Calculated Values for Units A, B, and C

Unit	θ_j	$\lambda_j \theta_j$
A	1.7	85
B	2.6	65
C	1.7	17
—	—	$\sum \lambda_j \theta_j = 167$

Table 4.18
Allocated Mean Active Corrective
Maintenance Time for System Units

System Unit	$\overline{T}_{acj}$
A	0.4326
B	0.6617
C	0.4326

Thus the allocated mean active corrective maintenance times for system units A, B, and C are 0.4326, 0.6617, and 0.4326 hours, respectively.

PROBLEMS

1. What is the difference between an FMEA and an FMECA?
2. Describe the basic steps used to perform an FMECA.
3. Discuss the following two techniques associated with an FMECA:
   - Risk priority number method
   - Military standard method

4. List the important advantages of an FMECA.
5. Comparisons between an FTA and an FMECA.
6. Describe the following two logic symbols associated with an FTA:
   - AND gate
   - OR gate
7. Assume that a windowless room has one light bulb and one switch. The switch never fails to open. Develop a fault tree for the event "Dark Room."
8. What are the benefits and drawbacks of performing an FTA?
9. Describe a cause and effect diagram and its associated basic steps.
10. Discuss the important elements of TQM.
11. What are the pitfalls related to TQM?
12. Define the term maintainability allocation.

REFERENCES

1. Bowles, J. B. and Bonnell, R. D. "Failure Mode, Effects, and Criticality Analysis," in *Tutorial Notes: Annual Reliability and Maintainability Symposium, 1994.* Evans Associates, Durham, North Carolina, 1994, pp. 1–34.
2. Countinho, J. S. "Failure-Effect Analysis." *Transactions of the New York Academy of Sciences,* Vol. 26, 1964, pp. 564–584.
3. Arnzen, H. E. *Failure Mode and Effects Analysis: A Powerful Engineering Tool for Component and System Optimization,* Report No. 347/40/00/00/K4-05 (C5776). The Government Industry Data Exchange Program, GIDEP Operations Center, Corona, California, 1966.
4. MIL-STD-1629, *Procedures for Performing a Failure Mode, Effects, and Criticality Analysis.* Department of Defense, Washington, D.C., 1979.
5. MIL-STD-1629A/Notice 2, *Procedures for Performing a Failure Mode, Effects, and Criticality Analysis.* Department of Defense, Washington, D.C., 1984.
6. Dhillon, B. S. "Failure Modes and Effects Analysis—Bibliography." *Microelectronics and Reliability,* Vol. 32, 1992, pp. 719–731.
7. *Instruction Manual, Potential Failure Mode and Effects Analysis in Design (Design FMEA) and for Manufacturing and Assembly Processes (Process FMEA).* The Ford Motor Company, Detroit, 1988.
8. Agarwala, A. S. "Shortcomings in MIL-STD-1629A Guidelines for Criticality Analysis," in *Proceedings of the Annual Reliability and Maintainability Symposium, 1990,* pp. 494–496.
9. Dhillon, B. S. *Systems Reliability, Maintainability and Management.* Petrocelli Books, Inc., New York, 1983.

10. Dhillon, B. S. and Singh, C. *Engineering Reliability: New Techniques and Applications.* John Wiley and Sons, New York, 1981.
11. Dhillon, B. S. and Singh, C. "Bibliography of Literature on Fault Trees." *Microelectronics and Reliability,* Vol. 17, 1978, pp. 501–503.
12. Grant Ireson, W., Coombs, C. F. and Moss, R. Y. *Handbook of Reliability Engineering and Management.* McGraw-Hill, New York, 1996.
13. Gevirtz, C. D. *Developing New Products with TQM.* McGraw-Hill, Inc., New York, 1994.
14. Dobyns, L. and Crawford-Mason, C. *Quality or Else.* Houghton Mifflin, Boston, 1991.
15. Schmidt, W. H. and Finnigan, J. P. *The Race Without a Finish Line: America's Quest for Total Quality.* Jossey-Bass Publishers, San Francisco, 1992.
16. Walton, M. *Deming Management at Work.* Putman, New York, 1990.
17. Van Ham, K. "Setting a Total Quality Management Strategy," in *Global Perspectives on Total Quality.* The Conference Board, New York, 1991.
18. Madu, C. N. and Chu-hua, K. "Strategic Total Quality Management (STQM)," in *Management of New Technologies for Global Competitiveness.* C. N. Madu, editor, Quorum Books, Westport, Connecticut, 1993, pp. 3–25.
19. Burati, J. L., Matthews, M. F. and Kalidindi, S. N. "Quality Management Organizations and Techniques," *Journal of Construction Engineering and Management,* Vol. 118, March 1992, pp. 112–128.
20. Imai, M. *Kaizen: The Key to Japan's Competitive Success.* Random House, Inc., New York, 1984.
21. Perisco, J. "Team Up for Quality Improvement." *Quality Progress,* Vol. 22, No. 1, 1989, pp. 33–37.
22. Coppola, A. "Total Quality Management," in *Tutorial Notes, Annual Reliability and Maintainability Symposium, 1992,* pp. 1–44.
23. Mears, P. *Quality Improvement Tools and Techniques.* McGraw-Hill, Inc., New York, 1995.
24. SAE G-11, *Reliability, Maintainability, and Supportability Guidebook.* Society of Automotive Engineers, Warrendale, Michigan, 1990.
25. Arsenault, J. E. and Roberts, J. A., editors. *Reliability and Maintainability of Electronic Systems.* Computer Science Press, Inc., Potomac, Maryland, 1980.
26. Chipchak, J. S. "A Practical Method of Maintainability Allocation." *IEEE Transactions on Aerospace and Electronic Systems,* Vol. 7, No. 4, 1971, pp. 585–589.

5

Specific Maintainability Design Considerations

INTRODUCTION

A cost effective and supportable design must take into account the maintainability considerations that arise at each phase in the life cycle of the system or product. It is therefore the role of maintainability engineering to ensure that those considerations, and the design factors related to them, receive full attention during the design process.

All efforts to establish maintenance concepts and requirements will be of little use if the design effort does not take into consideration the particular features that will enhance field maintenance. Thus careful planning and systematic effort are needed to bring attention to important maintainability design factors such as maintainability allocation, maintainability evaluation, maintainability design characteristics, maintainability parameters, and maintainability demonstration. Each of these factors involves various subfactors—for example, packaging, standardization, interchangeability, human factors, safety, and testing and checkout all play a role in the final product's maintainability design characteristics. Standardization, interchangeability, modular design, and accessibility are important considerations in every aspect of maintainability design.

MAINTAINABILITY DESIGN CHARACTERISTICS

These are the features and design characteristics that help reduce downtime and enhance availability. The goals of maintainability design include minimizing preventive and corrective maintenance tasks; increasing ease of maintenance; decreasing support costs; and reducing the logistical burden by decreasing the resources required for maintenance and support, such as spare parts, repair staff, and support equipment.

The most frequently addressed maintainability design factors, ranked in descending order, are: accessibility; test points; controls; labeling and coding; displays; manuals, checklists, charts and aids; test equipment; tools; connectors; cases, covers and doors; mounting and fasteners; handles; and safety factors. Other factors are standardization, modular design, interchangeability, ease of removal and replacement, indication and location of failures, illumination, lubrication, test adapters and test hookups, servicing equipment, adjustments and calibrations, installation, functional packaging, fuses and circuit breakers, cabling and wiring, weight, training requirements, skill requirements, required number of personnel, and work environment [1].

STANDARDIZATION

This important design feature restricts to a minimum the variety of parts and components that a product or system will need. Standardization can also be described as the attainment of maximum practical uniformity in a product's design [2, 3]. It represents the choice, design, or manufacture of parts, assemblies, equipment, associated tools, service materials, and/or procedures that are identical to or replaceable with other items or procedures. Standardization should be a central goal of design, because the use of non-standard parts may lead to lower reliability and increased maintenance. But while standardization is highly desirable, it cannot be allowed to interfere with advances in technology or improvements in design. Study of the advantages and disadvantages of standardization in this regard should therefore precede decision making. Lack of standardization is usually due to poor communication among design engineers, contractors, subcontractors, users, and other parties [4].

Standardization Goals

Some of the primary goals of standardization include maximizing the use of common parts in different products; minimizing the number of different types of parts, components, assemblies, and other items; maximizing the use of interchangeable and standard or off-the-shelf parts and components; minimizing the number of different models and makes of equipment in use; controlling and simplifying inventory and maintenance; reducing storage problems, and the effort spent on part coding and numbering.

Contributory Factors to a High Failure Rate for Nonstandard Items

Factors that contribute to the high failure rate for nonstandard items include [3]:

- Poor uniformity of manufacture in small production runs. Consequently, parts would need to be ordered more frequently, increasing the logistical burden. Also, small production runs could make the parts more difficult to obtain.
- Lengthy storage periods for items for which there is low demand and the resulting deterioration of the items during this time.
- Incorrect use, application, handling, installation, or maintenance of parts and components that are less familiar to workers.

Specific Applications of Standardization

There are many ways to make use of standard parts, components, and circuits. For example, equipment can contain interchangeable items such as batteries, controls, air cleaners, starting motors, and instruments. Regulators and supply voltages can be set to standard values. The materials, such as oils and fuels, used to service equipment can be made as uniform as possible. The advantages of standardization are shown in Figure 5-1.

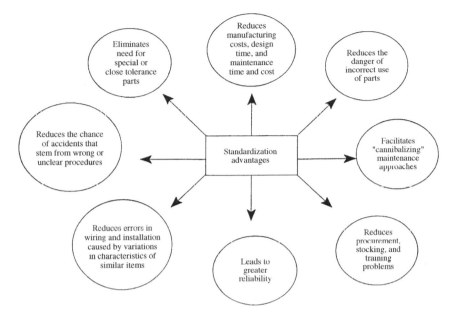

Figure 5-1. Benefits of standardization.

INTERCHANGEABILITY

This is an important maintainability design factor that is made possible through standardization. Interchangeability means that, as an intentional aspect of design, any component, part, or unit can be replaced within a given product or piece of equipment, by any similar component, part, or unit. There are two types of interchangeability, functional interchangeability and physical interchangeability. In functional interchangeability, two specified items serve the same function. In physical interchangeability, two items can be mounted, connected, and used effectively in the same locations and in the same manner. Interchangeability requirements should take into account field use conditions and the economy of manufacture and inspection.

Basic interchangeability principles include:

- Liberal tolerances.
- In equipment, products, or units that require frequent replacement and servicing of component parts due to wear or damage, each

and every part must be totally interchangeable with each and every other like part.

- In equipment, products, or units that are normally expected to operate without part replacement, strict interchangeability could be uneconomical.

The two broad categories of interchangeable parts are universally interchangeable parts and locally interchangeable parts. Universally interchangeable parts must be interchangeable in the field environment even if they have been produced by different facilities. Locally interchangeable parts are interchangeable with other like parts manufactured at the same facility but are not necessarily interchangeable with parts produced at different facilities. A design engineer must ensure the following to achieve maximum interchangeability of parts and units in a specified system [1]:

- When physical interchangeability is a design characteristic, there should also be functional interchangeability.
- When functional interchangeability is not desired, physical interchangeability is not necessary.
- Differences should be avoided in mounting, size, shape, and other such characteristics.
- Part and unit modifications should not change the methods of mounting and connecting. If this is not possible, incorporate them in a system or an assembly.
- There should be enough information available in job instructions and on plate identification so that the user can decide with confidence whether or not two like parts or units are interchangeable.
- All parts and units expected to be identical should be totally interchangeable and should be identified as interchangeable, carrying some manufacturer's number or other appropriate identification.
- If possible, there should be adapters provided to make physical interchangeability possible in situations where total—both functional and physical—interchangeability is not practicable.

Checklists for effectively incorporating interchangeability into equipment and product design contain questions such as [3]:

- Is there functional interchangeability whenever physical interchangeability is feasible?

- Is there total interchangeability wherever possible?
- Does complete interchangeability in fact exist for items that are expected to be identical and interchangeable?
- Does interchangeability exist for items with high failure rates?
- Are differences in mounting, size, and shape being avoided unless these differences serve a functional purpose?
- Are identical parts or components used wherever feasible in similar products or systems?
- Is sufficient and appropriate information provided on identification plates so that the user can easily judge whether like items are interchangeable?
- Are items such as parts, connectors, and cables standardized throughout the equipment in question?
- Are parts and units designed for functional interchangeability where complete interchangeability is not feasible?
- Are items such as screws and bolts the same size for all covers and cases?
- Does total electrical and mechanical interchangeability exist on all similar removable parts?
- Will mounting holes and brackets handle parts from different manufacturers?

MODULARIZATION

Modularization is the division of a system or product into physically and functionally distinct units to allow removal and replacement. Each system or subsystem, from the highest to the lowest level, can be designed as a removable entity. Questions of cost, practicality, and function dictate the degree of modularization. However, modular construction should be used wherever it is logistically feasible and practical, and will reduce training costs or provide other concrete benefit.

Modularization Design Guidelines

Some of the guidelines for designing modularized products are to [3, 5]:

- Divide the equipment into many modular units.
- Make modules and parts as uniform in size and shape as possible.

- Match the functional design of the equipment with division of the equipment into removable and replaceable units.
- Aim to design all equipment so that a single person can replace any malfunctioning component.
- Design control levers and linkages to permit easy disconnection from components, so that replacing components is a simpler process.
- Take an integrated approach to design—that is, consider the problems of component design, materials, and modularization simultaneously.
- Design the modules for greatest ease of operational testing when they are removed from the equipment.
- Strive to make each module capable of being inspected independently.
- Place emphasis on modularization for forward levels of maintenance to increase operational capability.
- Aim to make each modular unit small and light enough that a single person can handle and carry it without any difficulty.

Advantages of Modularization

Some of the many advantages associated with modularization are [3]:

- It is relatively easy to maintain a divisible configuration.
- Modular replacement in the field requires lower skill levels and fewer tools.
- Modularization can simplify new equipment design and shorten design time.
- Fully automated methods can be used to manufacture standard "building blocks."
- It is easier to divide up maintenance responsibilities in the most effective manner.
- Training maintenance staff becomes less costly and less time consuming.
- As recognition, isolation, and replacement of faulty items becomes easier, maintenance becomes more efficient and equipment downtime decreases.
- Existing equipment can be modified, with the latest functional units replacing their older equivalents.

Disposable Modules

Disposable modules are designed to be discarded rather than repaired once they fail. The use of disposable modules makes most sense when:

- Repair is either costly or impractical.
- There is a considerable and favorable difference between the cost of disposable and maintainable modules, and between the acquisition, storage, and supply costs for the two kinds of modules.
- The maintainable modules require significant expenditure in materials, tools, and labor time.
- The advantages of disposable modules, discussed in the following text, outweigh the disadvantages.

Disposable Module Design Requirements, Advantages, and Disadvantages

Requirements

Disposable modules must be designed, produced, and installed so that:

- Parts with long lives are not discarded because of the failure of parts with short lives.
- Expensive parts are not discarded because of the failure of less expensive parts.
- Disposable modules are encapsulated as much as possible.
- The modules carry a statement that they should be discarded at failure.
- The level of maintenance disposable modules may receive before being discarded is clearly defined.
- The tests to be performed on disposable modules, and the results that should be shown before the modules are discarded, are clearly defined.

The most important benefits and drawbacks of a disposal-at-failure design are:

Advantages

- Smaller, simpler, more durable modules with a more reliable design
- Simpler and more concise trouble-shooting approaches
- Reduction in required manpower, tools, facilities, and repair time
- Better interchangeability and standardization of modules
- Fewer types of spare parts required
- Improved reliability because of the sealing and potting methods

Disadvantages

- Increased inventory required because of need to have replacement modules on hand at all times
- Decreased performance and reliability of modules because of production efforts to keep them inexpensive to justify their disposal
- Decrease in available data on maintenance and failures, which would be used to help improve design
- Inability to redesign disposable modules
- High level of consumption of disposable modules, because of high rate of replacement, increased by cases of unnecessary replacement

SIMPLIFICATION

Probably the most difficult element of maintainability to achieve, but the most important, is simplification. Simplification should be the constant goal of design. Even a complex product or piece of equipment should appear simple and straightforward to the user. A good designer incorporates important functions of a product into the design itself and uses as few components as sound design practices will allow.

ACCESSIBILITY

Accessibility is the relative ease with which a part or piece of equipment can be reached for service, replacement, or repair. Lack of accessibility is an important maintainability problem and a frequent cause of ineffective maintenance. A U.S. Army handbook on maintainability stated in 1976 [1]: "Gaining access to equipment is probably second only to fault isolation as a time-consuming maintenance activity, and when automatic fault-isolation equipment becomes available, it unquestionably will be first."

Accessibility, it should be noted, does not automatically constitute maintainability. The fact that an item to be repaired can be readily accessed does not in itself guarantee overall ease and cost-effectiveness of maintenance.

Factors Affecting Accessibility

The factors that affect accessibility include |1, 3|:

- The item's location and environment
- Maintenance tasks to be carried out through the access opening
- Types of tools and accessories needed to perform the required tasks
- Clothing worn by the technical staff
- Visual needs of staff carrying out the tasks
- Specified time requirements for performing the tasks
- Work clearances necessary for performing the tasks
- Danger associated with use of the access opening
- Distance to be reached to access the item
- Packaging of items behind the access opening
- Mounting of items behind the access opening
- Frequency with which the access opening is entered

Access Location

The way a piece of equipment is installed governs in part the location of its maintenance access openings. The access opening(s) should occupy a face of the piece of equipment that will be accessible in the usual installation. The following are guidelines for designing and placing access openings [3]:

- Ensure that access openings will be accessible under normal installation of the equipment.
- Place access openings for maximum convenience in conducting the anticipated maintenance tasks.

- Ensure that the location of access openings permits direct access to the parts that will require maintenance.
- Ensure that the access openings occupy the same face as associated features such as control, test point, and displays.
- Ensure that the access openings are a safe distance from high voltage points or hazardous moving parts.
- Ensure that the lower edge of a restricted access opening is no less than 24 inches or its top edge no greater than 60 inches from the floor or work platform.
- Ensure that the location of accesses is in conformance with height of work stands and carts that will be frequently used.
- Ensure that heavy units can be pulled out instead of lifted out.

Access Opening Size

The access opening must be the proper size to allow a repair person to perform his or her tasks effectively. The factors that should determine the size of access openings include the size and shape of the internal objects to which access is required; the necessity of removing and replacing the objects through the openings; once access is gained, the movements of the human body required for actions such as turning, pushing, and pulling; and the size required for a repair person to enter partially or fully through the access opening. The last two factors are determined, respectively, by dynamic and static body measurements. Tables 5.1 and 5.2 present minimum access size requirements for one-handed tasks and minimum aperture dimensions for tasks requiring penetration of a normally clothed repair person's whole body, head, or shoulder [3].

Additional Guidelines Related to Access Opening Design

Other guidelines to consider in the design of access openings are to:

- Label each access opening with a unique number, letter, or other identifier.

Table 5.1
Minimum Access Size Requirements (in inches) for One-handed Tasks

Task	Dimensions in Inches for a Bare-handed Repair Person Wearing Regular Clothes
Inserting parts or components	Width = 4.5, Height = 1.75
Inserting empty hand through smallest square hole	3.5
Placing arm through access up to the elbow	Width = 4.5, Height = 4
Inserting hand through the smallest square hole, holding a screwdriver 8 inches long with a handle 1 inch in diameter	3.75
Inserting empty hand held flat	Width = 4.5, Height = 2.25
Inserting a closed hand with thumb outside of fist	Width = 5.125, Height = 4.25
Placing arm through access up to the shoulder, full arm's length	Width = 5, Height = 5

Table 5.2
Minimum Aperature Dimensions (in inches) for a Person Wearing Regular Clothes

Body Part or Position	Dimensions in Inches
Placing shoulders through access	Width = 20
Two standing persons passing through access side by side	Width = 36
Placing head through access	Width = 7
Person passing through access in kneeling position with back errect	Width = 20, Height = 64.5
Width of body passing through access	Width = 13
Person passing through access in crawling position	Width = 20, Height = 31

- In the case of small openings, indicate the position in which components or connectors should be inserted through the opening.
- Use safety interlocks on openings that lead to high voltage points.
- Round the edges of access openings.
- Identify on each access opening the items accessible through it.
- Furnish large access doors with a device to hold them securely open, because such doors might fall shut and cause damage.

- Provide efficient inspection apertures on items such as gear boxes and housings.
- Make the access opening that leads to the battery large enough to allow two-handed operation.
- Provide sufficient visibility to ensure safety for maintenance operations that involve hazard from nearby electrical circuits.
- When access openings are located near hazardous components, design the access door to that at its opening an internal light automatically indicates the danger points.
- Locate access openings to protect workers from contact with sharp edges, hot or moving parts, or other potential hazards.

IDENTIFICATION

Adequate labeling or marking of parts, controls, and test points facilitates maintenance tasks such as replacement and repair. If a repair person is unable to readily identify parts, test points, or controls, maintenance tasks become more difficult, take longer to perform, and are more likely to be performed incorrectly.

Types of identification include:

- Equipment identification: Equipment labeled, stamped, engraved, or otherwise marked with a permanent identifier.
- Instruction plates: Plates, bearing usage instructions, that are permanently attached to the item in a clearly visible location.
- Part identification: A schematic diagram that identifies each relevant part. The markings should be on or immediately adjacent to the referenced part and should be accurate and adequate to identify the part. They should last throughout the equipment's useful life and should be placed so that they can be easily read with the unit in its usual installed position and without the removal of any other parts. Table 5.3 presents appropriate letter and numeral sizes for a 28-inch viewing distance [6].

ACCESSIBILITY AND IDENTIFICATION CHECKLIST

Professionals working in the maintainability field have prepared a checklist on issues of accessibility and identification that includes the following points [3]:

Table 5.3
Appropriate Letter and Numeral Sizes for a 28-inch Viewing Distance

Type of Markings	Low Brightness (down to 0.03 ft-Lambert)	High Brightness (down to 1 ft-Lambert)
Critical markings in variable positions (e.g., numerals on counters and moving scales)	0.20–0.30 inches	0.12–0.20 inches
Noncritical markings (e.g., routine instructions, instrument identification labels)	0.05–0.20 inches	0.05–0.20 inches
Critical markings in fixed positions (e.g., numerals on fixed scales and switch markings)	0.15–0.30 inches	0.10–0.20 inches

- Windows allow users to inspect internal parts without physically entering the equipment.
- Consider using hinged doors where physical access is required.
- Determine the optimum placement of hinges.
- In the case of screw-fastened access places, try to use no more than four screws.
- Access openings should not have sharp edges.
- Label all access openings individually.
- Labels should indicate what parts can be reached through the openings.
- Label access points to specify the maintenance tasks that the parts accessible from this point will require and the anticipated frequency of the task.
- Access openings should provide sufficient space for the use of tools such as soldering irons and test probes.
- The access opening should also provide sufficient space for tasks requiring both arms and hands.
- Large parts that are difficult to remove can prevent access to other parts.
- In some designs, structural components prevent access to some parts.

- There should be maximum accessibility to parts needing regular inspection, replacement, or maintenance.
- Try to provide access to disposable modules that does not require the removal of other parts.
- In the design of devices that require lifting, pushing, or turning, take human strength limits into account.
- Units should be removable along a straight or moderately curved line.
- Heavy units, those weighing more than 25 pounds, should be located within the normal reach of repair persons.
- Make provisions for support of units during their removal or installation.
- If bolts, hoses, waveguides, or other items need to be removed before a maintenance or inspection task, they should be designed for easy removal.
- Layout should take into account the kinds of inspection that will need to be performed.
- Consider using split line design.
- In the design and location of all pieces of equipment that can be manipulated, consider relevant environmental factors.
- The access opening should provide adequate visibility for repair persons working within it.

GENERAL MAINTAINABILITY DESIGN GUIDELINES AND COMMON MAINTAINABILITY DESIGN ERRORS

Figure 5-2 shows some of the important general design guidelines that maintainability professionals have developed [7].

Many studies indicate that equipment designers have often made design errors that affect resulting equipment or system maintainability. Those design errors include the following [7]:

- Access doors containing many small screws.
- Access door handles not installed or incorrect handles installed.
- Adjustments placed out of reach.
- Adjusting screws that are difficult for maintenance personnel to locate.

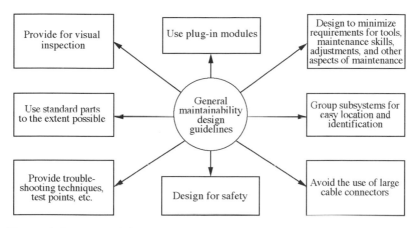

Figure 5-2. Some of the important general maintainability design guidelines.

- Adjusting screws installed too close to an exposed power supply terminal or to a hot component.
- Screwdriver-oriented adjustments located underneath modules in such a way that it is difficult for repair personnel to reach them.
- Low-reliability parts located underneath other parts requiring maintenance personnel to disassemble many other parts to reach them.
- Subassemblies that are screwed together in such a way that maintenance staff are unable to distinguish what is held by each screw.
- Insufficient room for a worker to make required adjustments when wearing a glove.
- Removable items installed in a manner that makes it unfeasible to remove them without disassembling the entire unit from its case or to remove other parts.
- Different modules designed with identical sockets and connectors, creating the risk that modules will be installed in the wrong place.
- Fragile parts placed just within the lower edge of the chassis, where it is more likely that they will accidentally be broken.
- The use of low-reliability test equipment that falsely reports product failures.

PROBLEMS

1. Discuss in detail at least five important design characteristics of maintainability.
2. What are the goals of standardization?
3. What are the advantages and disadvantages of standardization?
4. Discuss the important factors in achieving maximum interchangeability of parts within a given system.
5. Define the following:
   • Modularization
   • Interchangeability
   • Accessibility
6. List important advantages and disadvantages of a disposal-at-failure design.
7. List the factors that affect accessibility.
8. Describe three types of identification.
9. What are the common errors in maintainability design?

REFERENCES

1. AMCP 706-133, *Engineering Design Handbook: Maintainability Engineering Theory and Practice.* Department of Defense, Washington, D.C., 1976.
2. Ankenbrandt, F. L. et al. *Maintainability Design.* Engineering Publishers, Elizabeth, New Jersey, 1963.
3. AMCP 706-134, *Engineering Design Handbook: Maintainability Guide for Design.* Department of Defense, Washington, D.C., 1972.
4. Rigby, L. V. et al. *Guide to Integrated System Design for Maintainability,* Report No. ASD-TR-61-424. U.S. Air Force Systems Command, Wright-Patterson Air Force Base, Ohio, 1961.
5. Altman, J. W. et al. *Guide to Design of Mechanical Equipment for Maintainability,* Report No. ASD-TR-61-381. U.S. Air Force Systems Command, Wright-Patterson Air Force Base, Ohio, 1961.
6. TM 21-62, *Manual of Standard Practice for Human Factors in Military Vehicle Design,* Human Engineering Laboratories, Aberdeen Proving Ground, Maryland, 1962.
7. Pecht, M., editor. *Product Reliability, Maintainability, Supportability Handbook.* CRC Press, Boca Raton, Florida, 1995, pp. 191–192.

CHAPTER 6

Human Factors Considerations

INTRODUCTION

Human factors engineering goes back as far as 1898, when Frederick W. Taylor worked to design effective shovels [1], but it has been a key component of maintainability work only since World War II. The performance of military equipment during the war proved that equipment is only as good as the people operating and maintaining it. For the equipment to perform at its maximum capacity, operating and maintenance personnel must also perform effectively. But these personnel may fail to do so for a variety of reasons, such as inadequate attention to human factors during equipment design or poor training [2].

Human factors play an important role in both equipment reliability and equipment maintainability, but to a different degree. Reliability deals with those inherent characteristics of the equipment that promote either failure or long life. In comparison, maintainability has to do with the servicing, diagnostic, inspection, and repairability characteristics of the equipment. Maintainability is more dependent upon the action of the operating and maintenance personnel, and to a greater extent involves the interactions between people and machines. It therefore depends more on human factors than does reliability. Failure to consider human factors carefully during equipment design can lead to increased maintenance problems and reduced effectiveness and readiness.

HUMAN FACTORS PROBLEMS IN MAINTENANCE AND TYPICAL HUMAN BEHAVIORS

Human factors engineering applies knowledge about human capabilities, strength, and size to equipment design. Failure to effectively consider such factors can lead to serious maintainability and maintenance problems. For example, a study conducted by the United States military services revealed that over a 15-month period, human errors in aircraft maintenance contributed to 475 accidents and incidents in flight and ground operations [3]. The consequences were 96 seriously damaged or destroyed aircraft and the loss of 14 lives. Analysis of these accidents and incidents showed that many of them occurred shortly after periodic inspection. Furthermore, many of the human failures that caused them were repetitive failures. The basic causes of these errors were poor inspection; poor basic training in appropriate maintenance policies, procedures, and practices; poor supervision; and poor training in the maintenance of the specific equipment involved.

Because of these problems, equipment designers must minimize the likelihood of human error, and the consequences of potential errors, to the extent possible. For instance, they should reduce the number of support tasks required, design equipment so that the available personnel can easily accomplish the required support tasks in the given environment, and try to build in features that will make it impossible to perform required tasks incorrectly.

The designer should take into consideration typical human behaviors [4, 5]:

- People usually perform their assigned tasks while thinking about other things.
- People tend to use their hands for testing or examining.
- People are usually too impatient to take the necessary time to observe precautions.
- People usually read instructions and labels incorrectly or overlook them altogether.
- After performing a procedure, people usually fail to recheck their work for errors.
- People usually respond irrationally in emergency situations.
- People are usually reluctant to admit mistakes.

- After successfully handling hazardous objects over a lengthy period of time, many people become complacent and less careful.
- People are usually reluctant to admit that they do not see objects clearly, whether because of poor illumination or poor eyesight.
- People often estimate distance, speed, or clearance poorly. They usually underestimate large or horizontal distances and overestimate short distances.

HUMAN BODY MEASUREMENTS

This is important information in designing for maintainability. Designers must ensure in the early phases of their work that the equipment will accommodate operating and maintenance personnel of varying size, weight, and shape. There are two basic sources of information on body measurements:

- **Anthropometric surveys.** These surveys take measurements of a sample of the population and present the data in the form of percentiles, means, medians, and ranges.
- **Experiments.** These simulate the conditions of the equipment design being considered and allow measures to be taken under those conditions.

Designers can use data from either source or from both. The determining factor is usually the cost of performing experiments and the amount of satisfactory anthropometric survey data available.

Types of Body Measurements and Pointers for the Application of Force and the Strength of Body Parts

Equipment designers should take into account both static and dynamic body measurements. Static measurements represent the body in rigid standardized positions. They usually encompass more detail, including everything from overall body size to the distance between the eye pupils. Dynamic measurements represent the body in motion. Subjects assume varying work positions and allow functional leg and arm reaches to be determined. Dynamic measurements indicate how humans will perform a given task in a given space rather than how well they will fit into the space [2, 6].

Body Measurement Data for Use in Equipment Design

Table 6.1 gives dimensions for small and large men [7].

Weight Lifting Data

Maintenance tasks often require lifting various types of items. Designers should therefore give careful consideration to human weight-lifting capability under given situations when allocating tasks. Table 6.2 presents data concerning the maximum weight an adult male can lift from the ground. The figures come from a study of 19 males with average age, weight, and height of 21.6 years, 161.2 pounds, and 69.5 inches, respectively [8]. These individuals lifted, with no space limitations, objects of convenient size and shape.

The following are pointers for equipment designers concerning the application of body force and strength [9]:

- The maximum handgrip strength of a 25-year-old male is around 125 pounds.
- The use of the whole arm and shoulder increases the maximum exertable force. But using just the fingers requires the least amount of energy per degree of force applied.
- Factors such as body position, direction of force applied, body parts involved, and the object involved determine the degree of force that can be exerted.
- For side-to-side motion, push force is greater than pull force, and the maximum is approximately 90 pounds.
- Pull force is greater from a sitting than from a standing position. The maximum steady pull force is around 65 pounds, as opposed to the maximum momentary pull force of 250 pounds.
- Arm strength reaches its peak around 25 years of age and decreases slightly between the age of 30 years and 40 years. It declines approximately 40% from ages 30 to 65 years. During the same time frame, hand strength decreases roughly 16.5%.

Table 6.1
Body Measurement Data for Use in Equipment Design

Body Part	Measurement Description	Small Man (5th percentile, i.e., only 5% of the population is smaller than the value specified)	Large Man (95th percentile, i.e., only 5% of the population is larger than the value specified)
Hand	• Length	7"	8.2"
	• Width	3.2"	3.8"
Weight	(Without equipment)	130 lbs.	201 lbs.
Arm	Length from elbow to finger	17.3"	20.1"
Height	• Stature (standing)	65.5"	74"
	• Height of knees	21.0"	24.5"
	• Height when sitting erect	33.5"	38.0"
	• Seat height (popliteal height)	16.7"	19.2"
	• Eye height (internal canthus) in normal sitting position	28"	31.5"
	• Buttock-shoulder height (acromial height)	22.7"	26.5"
Head	• Length	7.2"	8.2"
	• Width	5.6"	6.4"
Foot (with shoe)	• Length	11.0"	12.7"
	• Width	4.0"	4.5"
Trunk	• Shoulder width	16.5"	20"
	• Seat width	13.0"	16.5"
	• Chest depth (front to back)	7.5"	11.0"
	• Width between outside of elbows (with arms hanging at sides)	15.3"	20.3"
Arm	• Arm span	65.9"	75.6"
	• Length of arm (functional anterior)	29.0"	35.0"
	• Elbow span (with arms spread out)	34.0"	39.0"

Table 6.2
Maximum Male Weight-lifting Capacity for Selected
Distances above Floor Level

Distance above Floor Level in Feet	Approximate Maximum Weight-lifting Capacity, in Pounds
1	143
2	140
3	76
4	55
5	40

HUMAN SENSORY CAPACITIES

Maintainability design also requires an understanding of human sensory capacities as they apply to areas such as color coding, shape coding, parts identification, and noise. The five major senses are touch, smell, sight, taste, and hearing. In addition, humans can also sense pressure, vibration, temperature, position, rotation and linear motion, and acceleration (or shock) [2, 3].

Sight

Electromagnetic radiations of specific wavelengths, generally known as the visible portion of the electromagnetic spectrum, stimulate sight. The human eye is more sensitive to certain colors at certain degrees of brightness than others. For example, it is most sensitive to greenish-yellow light with a wavelength of about 5500 Angstrom units. The eye also sees differently from different angles. Humans can perceive all colors while looking straight ahead, but color perception starts to decrease as the viewing angle increases. Colors disappear at certain degrees off the level view in the vertical plane [2, 9]:

- Green at 40°
- Red at 45°
- Blue at 80°
- Yellow at 95°
- White at 130°

They disappear at the following angles off the horizontal plane [2, 9]:

- Green-red at 60°
- Blue at 100°
- Yellow at 120°
- White at 180°

If equipment possesses color-banded meters, or warning lights of varying colors, at positions close to the horizontal or vertical limits of color differentiation, distinguishing the colors may be impossible. Designers should keep two other aspects of color perception in mind:

- **Color reversal.** This phenomenon can occur when, for example, a worker stares at a green or red light and then glances away. The signal to the brain may reverse the color.
- **Color perception under poorly illumination or nighttime conditions.** Under such conditions, it may be impossible to determine the color of a small point source of light—e.g., a small warning light or a light seen at a distance. In such cases, whether the light is orange, green, blue, or yellow, it will appear to be white.

Designers should also avoid placing too much reliance on color when an error in color perception could affect important operations; choose colors that will not confuse people who are color-blind or have other problems with color perception; and use, when possible, red filters with wavelengths greater than 6500 Angstrom. (If this is not possible, warning lights should be as close to red as possible.) [10].

Touch

As the maintenance of engineering equipment becomes a more complex and sophisticated task, it is important that maintenance personnel make effective use of all their senses. The sense of touch compliments human ability to interpret visual and auditory stimuli. For example, many control knob shapes can be easily recognized by touch alone. Such shapes would be useful in situations where the user is expected to rely entirely on his or her sense of touch.

Craft workers have for many centuries effectively used their sense of touch to detect surface irregularities and roughness. According to

S. Lederman [11] the detection of surface irregularities becomes dramatically more accurate when the worker moves a piece of paper or thin cloth over the surface rather than simply using his or her bare fingers.

Noise

Noise may be described as any undesirable sound. Human reaction to noise extends beyond auditory stimulation to include sensations such as irritability, fatigue, or boredom. Noise can adversely affect the effective performance of tasks that require intense concentration or a high degree of muscular coordination and precision. Excessive noise can also make oral communication difficult or impossible and can damage hearing. Two important characteristics of noise are [2, 3]:

- **Frequency.** This is measured in hertz (Hz) and the human ear can detect sounds ranging from 20 to 20,000 Hz. In particular, the ear is most sensitive to frequencies between 600 and 900 Hz. People exposed for long periods to noise at frequencies from 4,000 to 6,000 Hz normally suffer major hearing loss.
- **Intensity.** This is the major factor in the sensation of loudness and is usually measured in decibels (dB). Exposure to noise above 80 dB may lead to temporary or permanent hearing loss, but it is the length of exposure that determines the extent of damage.

Table 6.3 presents selected maximum background noise levels for reliable comprehension of words spoken in a raised voice by someone 36 inches to 48 inches away.

Table 6.3
Selective Maximum Noise Levels in dB
for Reliable Communication

Frequency (Hz)	Maximum Noise Level (dB)
Less than 75	79
75–150	73
150–300	68
300–600	64
4,800–10,000	57

Table 6.4 presents noise intensity levels in dB, and their effects, for various common sounds [3, 12].

Table 6.5 presents the durations of exposure to selected noise levels permitted by the Occupational Safety and Health Administration (OSHA) [12].

Table 6.4
Noise Intensity Levels in dB for Some Common Sounds

Noise Source	Intensity Level (dB)	Effect on People
Motion picture sound studio	10	Acceptable
A whispering voice	20	Acceptable
Quiet residential area	40	Acceptable
Household ventilating fan	56	Acceptable
Normal conversation	60	Acceptable
Heavy traffic	70	Acceptable
"Quiet" factory area	76	Acceptable
City bus	90	Reduced ability to hear may occur above this level
Punch press	110	Unacceptable and dangerous
Loud thunder	120	Unacceptable and dangerous
—	130	Approximate threshold of pain
—	150	Maximum permissible

Table 6.5
Duration of Exposure to Selected Noise Levels Permitted by the Occupational Safety and Health Administration

Noise Intensity Level (dB)	Exposure in Hours Per Day
105	1
100	2
97	3
95	4
90	8

To decrease the effects of noise [2]:

- In areas requiring the presence of maintenance personnel, keep noise levels below 85 dB.
- In areas that require maintenance actions in the presence of extreme noise, incorporate into the equipment appropriate acoustical design and mufflers and other soundproofing devices.
- Prevent unprotected repair workers from entering locations with sound levels greater than 150 dB, even for short periods.
- In situations where noise reduction is not possible, protect maintenance workers by issuing protective devices.

Vibration

The performance of both mental and physical tasks can be seriously affected by vibrations. In particular, large-amplitude, low-frequency vibrations contribute to problems such as headaches, fatigue, motion sickness, and eye strain, and interfere with the ability to read and interpret instruments effectively. All these symptoms become less pronounced as vibration amplitude decreases and vibration frequency increases. Equipment designers must make every reasonable attempt to eradicate the possibility that the equipment will generate problematic levels of vibration. Some recommendations for lowering the effects of vibrations are to [12]:

- Design equipment to resist vibrations and shocks by using devices such as shock absorbers, fluid couplings, and springs.
- When vital maintenance actions depend on reading digits or letters printed on equipment ensure that the equipment is not subject to vibrations, or at least not to vibrations greater than 0.08 millimeters in amplitude. In circumstances that do not permit vibration-free displays, consider increasing display size and/or illumination.
- Consider the positions that operating or maintenance personnel will assume in working on equipment subject to vibrations. Seated persons are usually most affected by vertical vibrations and prone persons by horizontal vibrations. Avoid vibrations of frequencies between 3 and 4 Hz, since this is the resonant frequency of a seated person's vertical trunk.

ENVIRONMENTAL FACTORS

Effective maintainability design must take into consideration the effects of the environment under which an individual is expected to perform his or her tasks. Factors such as temperature, illumination, humidity, and air circulation may seriously affect the ability of a repair person to work effectively. In general, environmental factors fall into the three categories shown in Figure 6-1: the working environment, the physical environment, and the human environment.

The working environment includes factors such as illumination, ventilation, duration of work, the arrangement of operating and maintenance work spaces, operational conditions, and time of day. Elements of the physical environment include noise, vibration, wind, temperature, pressure, humidity, toxic fumes, dust, and radiation. The human environment involves human psychological, physical, and physiological capabilities and limitations. The following sections discuss some of the important environmental factors related to maintainability.

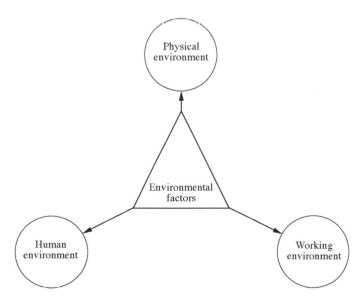

Figure 6-1. Environmental factors important to maintainability design.

Illumination

Light is radiant energy capable of exciting the retina of the eye and generating a visual sensation [12]. A repair person requires adequate illumination to perform his or her tasks effectively and without serious compromise of factors such as accuracy, speed, and safety. But because a satisfactory degree of illumination may not always be available, designers must ensure that maintenance tasks can be performed even under the poorest lighting conditions. For example, the equipment should be designed so that the maintenance tasks can be performed with only a flashlight, if that is all the illumination that will be available. Factors that should be carefully examined in designing any lighting system include:

• Appropriate level of brightness
• Uniformity of lighting
• Quality and color of illuminants and surfaces
• Glare from the lighting source or work surface
• Contrast in brightness between lighting for a given task and background light

Table 6.6 presents recommended illumination levels for various activities [3, 12].

Temperature

The effects of temperature on human performance are not entirely understood, but it is well known that certain temperature extremes decrease work efficiency. For example, as temperature increases above the comfort limit, motor responses and mental processes slow down and the likelihood of error increases. Similarly, as temperature falls below the comfort zone, physical fatigue and stiffening of the extremities begin.

The optimum temperature range for humans depends on various factors, but in general the recommended range is 65°F to 70°F. Other recommendations regarding heat are [2]:

• Install air conditioning, if at all feasible, when temperatures are likely to exceed 90°F.
• Install reflector/absorbant surfaces, to reflect or absorb heat, on equipment exposed to the sun.

Table 6.6
Recommended Illumination Levels for Various Typical Activities

Activity	Minimum Recommended Illumination Level in Fast-candles
Loading from a platform	2
Finding articles in bulk supply warehouse	5
Reading bulletin board	10
Filing records	25
Drafting	50
Repairing small components	100
Performing fine assembly	200
Performing extremely difficult inspection tasks	500
Performing surgical procedures	1000

- When equipment has components that require frequent maintenance located in excessively high temperature areas, redesign the equipment so that those components are located in cooler areas.

Designers should also carefully consider the following recommendations regarding cold:

- Structure tasks that must be performed in cold conditions to minimize sustained work time. For example, use quick-disconnect servicing equipment.
- Provide a large enough access opening and internal work area to allow repair persons to wear protective gloves when they must maintain equipment, such as liquid oxygen lines, that would be too cold for hands.
- Provide adequate heating in areas where maintenance personnel work.

Table 6.7 presents tolerable limits of heat and cold for people wearing normal clothing [13].

Relative Humidity

This is another environmental factor that affects human performance when combined with temperatures above or below the optimum range. In that range, humidity ranging from 30% to 70% is usually acceptable. At temperatures above this comfort zone, a comparatively small increase

Table 6.7
Tolerable Levels of Temperatures for People Wearing Normal Clothing

Description of Condition	Temperature (°F)
Comfortable during summer	56–75
Physical fatigue begins	75 and above
Deterioration of performance of mental or complex physical activities begins	85 and above
Comfortable during winter (light work)	63–71
Comfortable during winter (heavy work)	55–60
Physical stiffness of extremities begins	50 and lower

in relative humidity can have significant adverse effects on both the comfort and performance of maintenance personnel; as is well known, people can tolerate much hotter temperatures when air is dry rather than humid. Additional information on this subject is available in References 2, 3, and 14.

Air Circulation

In an enclosed work area, effective ventilation is essential for efficient performance. Usually, an adult needs 1,000 square feet of fresh air per hour while working. Also, for enclosed work spaces, the recommended rate of air circulation is 20 to 40 square feet per minute during cold weather and a little higher during hot weather.

Whenever there are or may be toxic materials in the air, ventilation alone is not adequate. Other measures to protect maintenance personnel must be taken. For example, the contamination source should be eradicated altogether or the involved personnel should be provided with protective devices. Additional information on the subject is available in References 2, 3, and 14.

AUDITORY AND VISUAL WARNING DEVICES

For the safety of maintenance personnel, a proper understanding of auditory warning devices is essential. Table 6.8 presents important characteristics of various types of alarms [2].

Table 6.8
Important Characteristics of Various Types of Alarms

Alarm Type	Ability to Get Attention	Frequency	Intensity	Ability to Penetrate Noise
Buzzer	Good	Low to medium	Low to medium	Fair, provided spectrum is suitable to noise in background
Siren	Very good, provided pitch rises and falls	Low to high	High	Very effective when frequency increases and decreases
Oscillator	Quite good, provided intermittent	Medium to high	Low to high	Quite good, provided frequency is selected properly
Bell	Good	Medium to high	Medium	Quite good when noise is low frequency
Whistle	Quite good, provided intermittent	Low to high	High	Quite good, provided frequency is selected properly

Auditory warning devices should also:

• Be easily detectable
• Be distinctive and both quickly and accurately identifiable
• Be suitable to hold a repair person's attention
• Use warbling or undulating tones and produce a sound at least 20 dB above the threshold level
• Not be continuous, high pitched tones above 2,000 Hz
• Not require interpretation while the maintenance person is conducting a repetitive task

Table 6.9 presents recommendations for designing for auditory alarm and warning devices.

Table 6.10 compares situations in which visual and auditory presentations may be used. Either auditory or visual presentations can be used when the signals will already be anticipated by the maintenance person, or when the signals will require discrete and short responses

Table 6.9
Design Recommendations for Auditory Alarm and Warning Devices

Condition to be Addressed	Design Recommendation
Signal must command maintenance person's attention	Modulate signal to generate intermittent "beeps"
Presence of background noise	Choose a frequency that makes the signal audible through other noise
Warning signal must be acknowledged	A manual shut-off mechanism will guarantee that the signal ends only when the required action is taken
Repair personnel are working far from the source of the signal	Make use of high intensities and avoid high frequencies
Sound is expected to pass through partitions and bend around obstacles	Make use of low frequencies

Table 6.10
Conditions for Using Visual and Auditory Presentations

Condition for Using Auditory Presentation	Condition for Using Visual Presentation
Simple message	Complex message
Message that requires immediate action	Message that does not require immediate action
Short message	Long message
Message receiving location that is too bright	Message receiving location that is too noisy
Maintenance person's job requires him/her to move about continuously	Maintenance person's job permits him/her to remain at one place
Message is associated with events in time	Message is associated with location in space
No reference will be made to the message later on	Reference will be made to the message at a later stage
Maintenance person overburdened with visual stimuli	Maintenance person overburdened with auditory stimuli

and will follow soon after another signal. The following three situations call for simultaneous use of auditory and visual signals:

- Warnings of extreme emergency
- Need for redundant signals
- Environmental conditions, such as high noise levels or poor illumination, that prevent data presentation through either visual or auditory means alone

SELECTED FORMULAS FOR HUMAN FACTORS

This section presents selected formulas that can be used by equipment designers as they take account of the role human factors play in maintainability.

Character Height

The success of many maintenance tasks depends on the repair person's ability to discern characters clearly. Usually, this ability depends on factors such as illumination, size of the characters, and exposure time. The following is a formula to determine character height by taking into consideration illumination, viewing distance, viewing conditions, and the importance of an accurate reading [15]:

$$H_C = kD_v + CF_{im} + CF_{iv} \qquad (6.1)$$

where k is the constant, with value 0.0022.

D_v is the viewing distance in inches.

CF_{iv} is the correction factor associated with illumination and viewing conditions: 0.06 (above 1 foot-candle, favourable reading conditions), 0.16 (above 1 foot-candle, unfavourable reading conditions), 0.16 (below 1 foot-candle, favourable reading conditions), 0.26 (below 1 foot-candle, unfavourable reading conditions).

CF_{im} is the correction factor associated with importance. For example, for important items such as emergency labels its recommended value is 0.075, and for others CF_{im} is set equal to zero.

H_C is the character height in inches.

Example 6-1

The estimated viewing distance of an instrument panel is 40 inches. Determine the height of the characters that should be used on the panel, if $CF_{iv} = 0.26$ and $CF_{im} = 0.075$.

Inserting the specified values into Equation 6.1 yields

$$H_C = (0.0022)(40) + 0.075 + 0.26 = 0.4230 \text{ in}$$

That is, the height of each character should be 0.4230 inches.

Lifting Load

Information on the maximum lifting load for an individual is useful in structuring maintenance tasks. This load is expressed by the formula [16]:

$$LL_m = CBMS_i \qquad (6.2)$$

where LL_m is the maximum lifting load for an individual.

C is the constant with value 1.1 for males and 0.95 for females.

BMS_i is the individuals's isometric back muscle strength.

The Decibel

This basic unit of noise/sound intensity is named after Alexander Graham Bell, inventor of the telephone. Sound-pressure level is defined by [12, 17]:

$$SPL = 10 \log_{10}(P^2/P_0^2) \qquad (6.3)$$

where SPL is the sound-pressure level expressed in decibels.

P_0^2 is the standard reference sound pressure, representing zero decibels, squared. Under normal circumstances, P_0 is the faintest 1000 Hz tone that an average young person can hear.

P^2 is the sound pressure, squared, of the sound to be measured.

Human Reliability

The reliability of repair personnel performing time-continuous tasks can be calculated using the following equation [18]:

$$R_m(t) = e^{-\int_0^t \lambda_m(t)dt} \qquad (6.4)$$

where $R_m(t)$ is the maintenance person's reliability at time t.
$\lambda_m(t)$ is the time-dependent error rate of the maintenance person.

Example 6-2

A maintenance person performs certain time-continuous tasks and his or her associated error rate, λ_m, is 0.0002 errors per hour. The time to error is exponentially distributed. Determine the reliability of the maintenance person for performing an assigned task correctly for a five-hour period.

In this case, we have

$t \quad = 5$ hours

$\lambda_m = 0.0002$ error/hour

Substituting the above values into Equation 6.4 yields

$$R_m(5) = e^{-\int_0^5 (0.0002)dt} = e^{-(0.0002)(5)} = 0.999$$

Thus, the reliability of the maintenance person to perform the assigned task correctly is 0.999.

PROBLEMS

1. Define the following two terms:
   - Human factors
   - Human factors engineering

2. What are the typical human behaviors that an equipment designer should take into consideration when assigning tasks to maintenance persons?
3. Discuss the following three human sensory capabilities:
   • Sight
   • Touch
   • Hearing
4. Discuss the effect of vibrations on the human body and human performance.
5. Describe the effects of illumination and temperature on human performance.
6. Discuss the important characteristics of siren and oscillator alarms.
7. List at least four design recommendations for auditory alarm and warning devices.
8. Describe the significance of the following:
   • Relative humidity
   • Air circulation
9. List at least seven conditions under which auditory presentations would best convey a message to maintenance personnel.
10. A repair person conducts certain time-continuous tasks and his or her associated error rate is 0.004 errors per hour. Calculate the reliability of the repair person for performing a task correctly for a seven-hour period. Discuss any assumptions made in the calculation.

REFERENCES

1. Chapanis, A. *Man-Machine Engineering.* Wadsworth Publishing, Belmont, California, 1965.
2. AMCP 706-134, *Engineering Design Handbook: Maintainability Guide for Design.* Department of Defense, Washington, D.C., 1972.
3. AMCP 706-133, *Engineering Design Handbook: Maintainability Engineering Theory and Practice.* Department of Defense, Washington, D.C., 1976.
4. Nertney, R. J. and Bullock, M. G. *Human Factors in Design,* Report No. ERDA-76-45-2. The Energy Research and Development Administration, U.S. Department of Energy, Washington, D.C., 1976.
5. Dhillon, B. S. *Engineering Design: A Modern Approach.* Richard D. Irwin, Inc., Chicago, 1996.

6. Morgan, C. T., editor. *Human Engineering Guide to Equipment Design.* McGraw-Hill Book Company, Inc., New York, 1963.

7. TM 21-62, *Manual for Standard Practice for Human Factors in Military Vehicle Design.* Human Engineering Laboratories, Aberdeen Proving Ground, Maryland, 1962.

8. Altman, J. W. *Guide to Design of Electronic Equipment for Maintainability,* Report No. WADC-TR-56-218. Wright Air Development Center, Wright-Patterson Air Force Base, Ohio, 1956.

9. Henney, K., editor. *Reliability Factors for Ground Electronic Equipment.* The Rome Air Development Center, Griffiss Air Force Base, Rome, New York, 1955.

10. Woodson, W. "Human Engineering Suggestions for Designers of Electronic Equipment," in *NEL Reliability Design Handbook.* U.S. Naval Electronics Laboratory, San Diego, California, 1955, pp. 12-1 to 12-5.

11. Lederman, S. "Heightening Tactile Impressions of Surface Texture," in *Active Touch.* G. Gordon, editor, Pergamon Press, New York, 1978.

12. McCormick, E. J. and Sanders, M. S. *Human Factors in Engineering and Design.* McGraw-Hill Book Company, New York, 1982.

13. Altman, J. W. *Guide to Design of Mechanical Equipment for Maintainability,* Report No. ASD-TR-61-381. U.S. Air Force Systems Command, Wright-Patterson Air Force Base, Ohio, 1961.

14. Woodson, W. E. and Conover, D. W. *Human Engineering Guide for Equipment Designers.* University of California Press, Berkeley, California, 1966.

15. Peters, G. A. and Adams, B. B. "Three Criteria for Readable Panel Markings." *Product Engineering,* Vol. 30, No. 21, 1959, pp. 55–57.

16. Poulsen, E. and Horgensen, K. "Back Muscle Strength, Lifting and Stooped Working Postures." *Applied Ergonomics,* Vol. 2, 1971, pp. 133–137.

17. Adams, J. A. *Human Factors Engineering.* Macmillan Publishing Company, New York, 1989.

18. Dhillon, B. S. *Human Reliability: With Human Factors.* Pergamon Press, Inc., New York, 1986.

Safety
Considerations

INTRODUCTION

Safety means either freedom from hazards or protection against hazards. It is one of the most important factors in designing for maintainability. As individuals perform maintenance tasks, they are exposed to hazards and accidents. Many of these hazards and accidents are due to careless design or design that does not give adequate attention to human factors and safety features. Other factors include hazardous environmental conditions and the creation of hazards by maintenance and operating personnel themselves when they perform their assigned tasks carelessly. The key to overcoming many of these difficulties is to "design in" safety features that will protect operators, maintenance personnel, and the equipment itself.

Past experience indicates that absolute safety is not attainable. Not all hazards can be designed out of equipment, and operating and maintenance personnel cannot be relied upon to observe safety precautions at all times. Furthermore, as Murphy's Law points out, if there is an incorrect way to do something, eventually someone will do it that way [1, 2]. Nonetheless, designers must remember that people are the single most valuable commodity and that equipment that is dangerous to people is by definition not maintainable.

SAFETY AND MAINTAINABILITY DESIGN

Over the years professionals working in maintainability design have developed guidelines to provide for the safety of operating and maintenance personnel. Some of those guidelines are [1]:

- Ensure that access openings are fitted with fillets and are large enough to allow easy entrance.
- Thoroughly study potential sources of injury by electrical shock.
- Locate items that will require maintenance so that personnel can gain access to them without danger from moving parts, heat, electric current, toxic chemicals, or other sources of hazard.
- Provide fail-safe devices so that a failure in one unit cannot lead to the failure of other units that would consequently result in serious damage to the system and possible human injury.
- Place emergency doors and other emergency exits so that they provide maximum accessibility.
- Provide guides, stops, and tracks to facilitate the handling of equipment.
- Provide sufficient fire-extinguishing and other equipment in areas where fire hazards are a possibility.
- Indicate weight-lifting or weight-holding capacity on items such as hoists, lifts, and jacks.
- Provide eye baths, showers, and other special equipment in areas where toxic materials will be handled.

ELECTRICAL, MECHANICAL, AND OTHER HAZARDS

Electrical shock is probably the major area of concern as even a small shock can be dangerous. Any potential for exposure to more than 50 volts should be categorized as a possible electric shock hazard. Studies indicate that contact with voltages ranging from 70 to 500 volts is the major cause of deaths from workplace accidents [3, 4]. However, the effect of electric shock depends on many factors: body resistance; the duration of the shock; current, voltage, and frequency;

and the physical condition of the individual. Duration is an important concern, because short electric shocks can trigger a heart attack as shown in Table 7.1 [5].

Maintenance personnel are probably more vulnerable to electric shock than any other group of workers because of the very nature of their duties. Thus, maintainability design must treat the possibility of electric shock as a central concern.

Methods to provide adequate protection for personnel include enclosing parts that might otherwise pose hazards, providing access-door safety switches, automatic operation of the main equipment switch at the instant of the door opening, and automatic component grounding in the event the unit is opened for access to the components [6].

- **Safety switches.** Used to prevent electric shock, these switches are either interlocks, battle-short switches, or main power switches. An interlock is a switch that automatically opens the power circuit of a piece of equipment at the instant an access door or cover opens. Interlock switches remove power during maintenance work, and they should be installed on each door and cover providing access to voltage in excess of 40 volts. The main purpose of a battle-short switch is to render all interlock switches inoperative, by placing a short circuit across all interlocks. The main power switch removes all power from the equipment by opening all connections from the main power source.
- **Discharging devices.** These devices are used in situations where high-grade filter capacitors can store lethal charges over a relatively long period. Use of discharging devices is recommended when the

Table 7.1
Electric Shock Durations that May Lead to Heart Attacks

Shock Duration in Seconds	Alternating Current (AC) in Milliamps (mA) and at 60 Cycles per Second	Direct Current (DC) in Milliamps (mA)
0.03	1,000	1,300
3.0	100	500

time constant of capacitors and associated circuits is greater than five seconds. The discharging devices must be reliable and must operate automatically at the instant the enclosure opens. Without the appropriately placed discharging devices, the safety of maintenance personnel and others could be compromised.

- **Safety markings.** These warn people of hazardous conditions and indicate what precautions they should observe to ensure their safety and that of the equipment. However, because safety markings or signs are not physical barriers to hazard, they should only be used when no other method of protection is feasible.
- **Safety warning devices.** Some examples are bells, lights, and vibration devices. Safety warning devices are installed at locations where maintenance personnel and other individuals can readily and clearly sense them.
- **Grounding.** There are many grounding methods used to protect individuals from dangerous voltages. Usually, exposed parts, chassis, and enclosures are grounded, using the same ground connector. A terminal spot welded to the chassis normally provides a reliable ground connector. But riveted elements should be avoided for grounding because of their poor electrical connection reliability.
- **Fusing.** This is another important method of safeguarding personnel from electric shock. Fuses are used to protect all leads from the primary service lines and should be connected to the lead side of the main power switch only.

During equipment design, potential mechanical hazards should also be eradicated to the extent possible. Guidelines for minimizing or eliminating mechanical hazards include: round all edges and corners to the largest radii practical; use flathead screws whenever possible; cover, coat, or machine smooth all exposed surfaces; use recessed mountings for small projecting parts, e.g., toggle switches and small knobs located on front panels; use shields and guards to prevent unintentional contact with rotating or oscillating parts such as gears, levers, and couplings; enclose or otherwise guard high-temperature parts; provide appropriate ventilation to keep component temperatures within specified limits; and avoid locating air exhaust openings on front panels [6].

Other significant hazards are:

- **Fire.** Every reasonable step should be taken to minimize or eliminate fire hazards during equipment design. In particular, items such as inductors, capacitors, or motors that pose potential fire hazards should be enclosed by a noncombustible material with the minimum possible number of openings. Also avoid designing equipment that will emit flammable vapors during operation or storage; ensure that equipment will not produce dangerous smoke or fumes; and provide hand-operated, portable fire extinguishers at appropriate locations.
- **Toxic fumes.** As with fire hazards, every step should be taken during design to eliminate hazards from toxic fumes. To give some examples of these hazards, exhaust from internal combustion engines contains numerous hazardous substances, and diesel engines emit aldehydes and nitrogen oxides. Table 7.2 presents a list of selected common toxic agents and their corresponding

Table 7.2
Selected Common Toxic Agents and Their Corresponding Maximum Allowable Concentrations

Toxic Agent	Source	Maximum Allowable Concentration (ppm)
Carbon dioxide	Engine exhaust	5,000
Carbon monoxide	Engine exhaust	100
Sulphur dioxide	Engine exhaust	5
Gasoline	Fuels and propellants	250
Ammonia	Fuels and propellants	100
Kerosene	Fuels and propellants	500
Nitrogen tetroxide	Fuels and propellants	5
Phosgene	Smoke	1
Aldehydes (acrolein)	Oil sprays and fumes	0.5
Aldehydes (furfural)	Oil sprays and fumes	5
Aldehydes (acetaldehyde)	Oil sprays and fumes	200
Carbon dioxide	Fire extinguishants	5,000
Methyl bromide	Fire extinguishants	20
Diacetone	Hydraulic fluids	50
Dioxane alcohol	Hydraulic fluids	100
Freon	Refrigerants	1,000
Methyl bromide	Refrigerants	20

maximum allowable concentrations [6, 7]. Maintainability professionals must take these risks into consideration.

- **Instability.** Instability is a proven cause of accidents. Thus, it is important to design all equipment for maximum stability. In particular, strive to maximize safety and stability in the design of equipment that will move up an incline such as a cargo ramp, or that will be lifted by cranes for shipping. Under such circumstances it is advisable to mark the equipment's center of gravity and jacking points.

- **Explosion.** Explosions are a common cause of fatalities. Equipment for use in an explosive environment must be designed so that all possibilities of explosion are eliminated. Any electrical equipment being designed for use in the vicinity of flammable gases or vapors must be explosion-proof. Every effort must be made to eliminate or minimize any danger to people by separating hazardous substances from heat sources and by incorporating spark arrestors, suitable vents and drains, and other safety features.

SAFETY ANALYSIS TOOLS

During the equipment design phase, professionals have many methods at their disposal for eradicating or minimizing hazards to people and equipment. The basic objective of all these approaches is hazard identification and control. System safety analysis should demonstrate that safety specifications, criteria, or objectives are fully satisfied; provide equipment designers with information for trade-off studies to achieve optimum system safety; and assist safety professionals to better understand, during equipment design, operation, and maintenance, potential safety problems.

Some of system safety analysis techniques are hazard analysis, preliminary safety matrix analysis, failure mode and effects analysis, fault tree analysis, and operating hazard analysis [8]. Three of these techniques are described in the following section.

Hazard Analysis Method

This powerful tool determines the safety requirements for people, procedures, and equipment used in testing, operations, maintenance, and logistic support. This method also determines the compliance of system and equipment with specified safety requirements and criteria. Hazard analysis is conducted in terms of the following levels of hazard [9].

- **Catastrophic.** This indicates that human error, design characteristics, procedural deficiencies, environment, or part/equipment malfunction will lead to fatality, severe injury, or system loss.
- **Critical.** This indicates that human error, design characteristics, procedural deficiencies, environment, or part/equipment malfunction will lead to human injury or major system damage or will require urgent measures to ensure the survival of people or the system.
- **Marginal.** This indicates that human error, design characteristics, procedural deficiencies, environment, or part/equipment malfunction can reliably be counteracted or put under control before human injury or major system damage occur.
- **Negligible.** This indicates that human error, design characteristics, procedural deficiencies, environment, or part/equipment malfunction that could lead to human injuries or major system damage will not occur.

As Figure 7-1 shows, hazard analysis involves identifying, and classifying the level of hazards, and highlighting of areas that will require special design attention to reduce or eliminate identified potential hazards, especially those classifiable as critical or catastrophic.

Hazard analysis should address areas such as material compatibility; relevant environmental constraints; isolation of energy sources; crash safety; resistance to shock damage; toxic fumes; fail-safe design methods; fuels and propellants and their associated problems; hazards associated with the use of explosive devices; danger of implosion; protective clothing or equipment required; effects of electromagnetic radiation, transient current, or electrostatic discharges from the equipment or on it; nuclear radiation and its effects; training needed to ensure safe equipment operation and maintenance; life support requirements' safety implications in manned systems; potential sources of fire ignition and propagation; the risk of human error in operation of the equipment; and questions about vulnerability to certain kinds of accidents and the level of danger those accidents would pose to people and equipment.

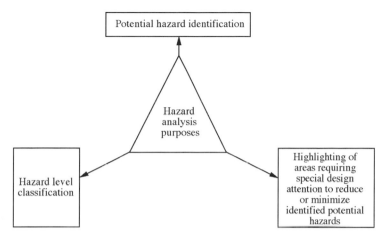

Figure 7-1. Basic purposes of hazard analysis.

Failure Mode and Effects Analysis

As discussed in Chapter 4, this is a widely used reliability analysis technique that also applies to maintainability analysis. The technique systematically determines the basic causes of failure and defines measures to reduce their effects. Furthermore, it can be applied to any system level. The failure mode is the specific way in which the item fails to carry out its intended mission. The failure cause is the reason the failure took place. The failure effect is the result of the failure for each failure mode.

- **Failure mode.** Examples are open or short circuits; reduced output, loss of function, and loss of output.
- **Failure cause.** Examples are wear, vibration, contamination, and voltage surge.
- **Failure effect.** Examples are loss of communication, mission abort, reduced control, and injury or damage to personnel or equipment.

FMEA is basically a qualitative approach to determining the reliability, maintainability, and safety of a given design by taking into consideration potential failures and their resulting effects. The seven major steps in performing FMEA are [10]:

- Define the boundaries and detailed requirements for the system or piece of equipment under consideration.
- List all of its components.
- List all possible failure modes, describe each, and identify the component that would be involved.
- Assign a failure rate to each component failure mode.
- List the effects of each failure mode.
- Enter remarks for each failure mode.
- Review each critical failure mode and take appropriate action.

In using this analysis, the effect identified can be quite different depending on the objective of the analysis. For example:

- **In reliability analysis.** The effect considered is the effect on the system's or equipment's performance or ability to function.
- **In maintainability analysis.** The effects considered include the symptoms through which failure to be pinpointed and the components that will require replacement as the result of the failure.
- **In safety analysis.** The effects considered are damage to other systems and equipment and possible danger to people.

Some of the advantages of the FMEA method are that it employs a systematic procedure to categorize hardware failure and identifies all possible failure modes and their effects on performance, personnel, and equipment. It is useful for comparing design, simple to understand, and helps identify methods of detecting the various possible failures.

Fault Tree Analysis

As discussed in Chapter 4, this is a widely used technique to determine reliability and safety by estimating the probability that an undesirable event will happen [10].

Fault trees provide a tool for accident analysis, a display of results, and a convenient and efficient format for describing reliability problems. In the case of accident analysis, a fault tree helps in finding all possible causes of an accident. If used effectively, the fault tree method often results in the discovery of failure combinations that otherwise might go unrecognized. Similarly, the display of results is useful when a design is unsatisfactory. The display shows weak points and the way they can lead to the undesired event. On the other hand, if the design is satisfactory, the fault tree helps to show that all possible causes of failure have been adequately considered. Fault tree analysis involves the following steps [11]:

- Selection of an undesired event or fault to be investigated. Examples might be injury to a worker, equipment failure, or inadvertent launch of a missile.
- Analysis of the functional flow diagrams for the equipment or system, to determine combinations of events and failures that could lead to an occurrence of the fault event.
- Development of suitable mathematical expressions, using Boolean algebra.
- Determination of circumstances under which each event associated with the fault tree could occur.
- Systematic diagramming of contributory events and failures to show their relationship to each other as well as to the undesirable event under investigation.
- Estimation of the failure rate of each item being considered or estimation of the occurrence probability of each event.
- Determination of the occurrence probability of the undesirable event being examined, based on the probabilities of occurrence of the contributory events.

There are numerous symbols used to construct a fault tree. Figure 7-2 presents five of the most commonly used symbols.

- **OR gate.** As described in Chapter 4, this represents a condition in which an output event occurs if any one or more of the n input events occur.
- **AND gate.** As described in Chapter 4, this represents a condition in which an output event occurs only if all of the n input events occur.
- **Resultant event.** This is denoted by the rectangle shown in Figure 7-2(c) and represents a condition in which an event is a result of the combination of fault events that precede it.
- **Basic event.** This is denoted by the circle shown in Figure 7-2(d) and represents the failure of an elementary component or a basic fault event.

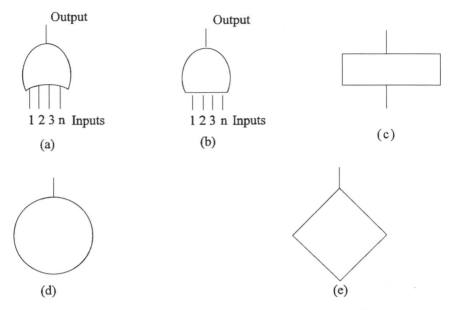

Figure 7-2. Commonly used symbols to construct a fault tree: (a) an OR gate, (b) an AND gate, (c) a resultant event, (d) a basic event, (e) an incomplete event.

- **Incomplete event.** This is denoted by the diamond shown in Figure 7-2(e) and represents a fault event whose cause has not been fully determined either due to lack of interest or due to lack of data.

Figure 7-3 demonstrates a simple fault tree analysis of an undesired event.

Some of the benefits and drawbacks of the fault tree analysis method are [10]:

Benefits of Fault Tree Analysis

- It is a tool that designers, management, and users can use to analyze failures and potential failures in visual terms.
- It ferrets out failures deductively.
- It provides insight into the behavior of the system or equipment.

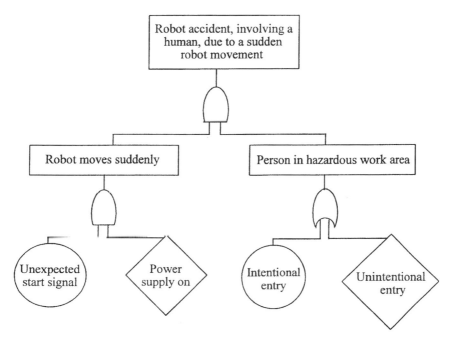

Figure 7-3. A simple fault tree for an undesired or top event: robot accident, involving a person, due to a sudden robot movement.

- It provides options for conducting qualitative and quantitative analysis.
- It forces reliability, maintainability, and safety analysts to understand the system or equipment under consideration thoroughly.

Drawbacks of Fault Tree Analysis

- The end result is difficult to verify.
- It can be a costly and time-consuming approach.
- It has difficulty handling states of partial failure.

Comparison of FMEA and Fault Tree Analysis

Table 7.3 presents a comparison of FMEA and fault tree analysis [1, 12].

Table 7.3
Comparison of FMEA and FTA

FMEA	FTA
It is a hardware oriented method.	It is an event oriented approach.
It has a broader scope with restricted depth of analysis.	It has a restricted scope with in-depth analysis.
It is an optimum approach for multiple failures.	It is an optimum approach for single failures.
It does not require analysis of failures that have no effect on the operation under investigation.	It provides documentation to ensure that each and every potential single failure has been investigated.
It does not require investigation of all external influences.	It highlights all external influences contributing to loss, for example, environment, test procedures, and human errors.

SAFETY AND HUMAN BEHAVIOR

The safety of the people who operate and maintain equipment is of utmost importance. During the design phase, appropriate information on human behavior can ultimately lead to safer and more maintainable design. Chapter 6 lists some typical human behaviors that can lead to injuries. Other examples are the following [13–16]:

- People often fail to look where they place their hands and feet, particularly in familiar surroundings.
- People often continue to use items or equipment they know to be faulty.
- People often underestimate the probability of occurrence of the "unpleasant event" and overestimate the probability of occurrence of the "pleasant event."
- People frequently underestimate cold temperatures and overestimate hot temperatures.
- People often underestimate compact weight and overestimate bulky weight.

- People frequently underestimate looking-up height and over-estimate looking-down height.
- People usually underestimate a decelerating object's speed and overestimate an accelerating object's speed.

Figure 7-4 shows important measures for reducing accidents caused by human error.

SAFETY CHECKLIST

There are many safety issues that must be carefully considered during the design process. The following list, which incorporates many points made in previous sections of this book, will help ensure that these issues are considered and that any necessary corrective measures are taken. Note that there may well be issues not on this list that should also be considered [6]:

- Mechanical guards on all moving equipment parts
- Critical warning lights
- Placement of fire extinguishers
- Selection of warning lights
- Location of dangerous voltage point
- Existence of warning plates at appropriate places

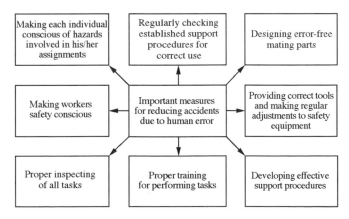

Figure 7-4. Measures for reducing accidents caused by human error.

- Use of on-off or fail-safe circuits
- Use of safety interlocks
- Rounded edges on components and maintenance access openings
- Distinctiveness of audible warning signals
- Clear identification of jacking and hoisting points
- Use of color code techniques
- Use of limit stops on drawers or fold-out assemblies
- Mounting of conspicuous placards adjacent to high-voltage or very hot equipment
- Existence of warning signals to indicate dangerous concentration of dangerous gases
- Inadvertent dislodging of heavy springs
- Location of adjustment screws and commonly replaced parts with respect to high voltage or hot parts
- Existence of bleeding devices for high-energy capacitors
- Design of fault detection systems
- Suitable auditory warning backup for displays that require continuous monitoring but that may not be watched continuously
- Strict separation of control circuits and warning circuits
- Safety of tools and equipment to be used in explosive environment
- Location of components and live wires that retain dangerous voltages even when the equipment is turned off
- Design of exhaust system
- Existence of struts and latches to secure hinged and sliding components against accidental movement
- Location and mounting of components with respect to access and safety
- Transparent window or removable cover installation over fuses
- Existence of portable hand-operated fire extinguishers at appropriate places
- Effective labelling or coding of all liquid, gas, and steam pipelines.
- Open-position positive locks for hatches
- Provisions to prevent sparking and other conditions associated with vehicle electrical systems
- Upward pointing of exhaust pipes of internal combustion engines
- Installation of warning lights to indicate fire or excessive heat in areas that are partially visible to drivers of vehicles
- Use of materials that might create hazards under extreme operating conditions

PROBLEMS

1. Describe the following four methods of electric shock prevention:
   - Safety switches
   - Safety warning devices
   - Discharging devices
   - Grounding
2. Discuss the following hazards:
   - Toxic fumes
   - Fire
   - Instability
3. Describe the method called hazard analysis.
4. Discuss the steps involved in performing FMEA.
5. What are the advantages and disadvantages of the fault tree analysis method?
6. Compare the FMEA and FTA approaches.
7. Describe the following symbols associated with the fault tree analysis method:
   - Rectangle
   - Circle
   - Diamond
8. Describe five typical human behaviors that could lead to injuries from unsafe acts.
9. List six important measures for reducing accidents caused by human error.
10. Discuss general safety guidelines associated with maintainability design.

REFERENCES

1. AMCP 706-133, *Engineering Design Handbook: Maintainability Engineering Theory and Practice.* Department of Defense, Washington, D.C., 1976.
2. NORD OD-39223, *Maintainability Engineering Handbook.* Department of Defense, Washington, D.C., 1970.
3. OP-2230, *Workmanship and Design Practices for Electronic Equipment.* Department of Defense, Washington, D.C., 1959.
4. Neibel, B. W. *Engineering Maintenance Management.* Marcel Dekker, Inc., New York, 1994.

5. TM 21-62, *Manual of Standard Practice for Human Factors in Military Vehicle Design.* Human Engineering Laboratories, Aberdeen Proving Ground, Maryland, 1962.
6. AMCP-706-134, *Engineering Design Handbook: Maintainability Guide for Design.* Department of Defense, Washington, D.C., 1972.
7. Henny, K., Lopatin, I., Zimmer, E. T., Adler, L. K. and Naresky, J. J. *Reliability Factors of Ground Electronic Equipment.* McGraw-Hill Book Company, New York, 1956.
8. Dhillon, B. S. *Reliability Engineering in Systems Design and Operation.* Van Nostrand Reinhold Company, New York, 1983.
9. MIL-STD-882, *System Safety Program for System and Associated Subsystem and Equipment—Requirements for.* Department of Defense, Washington, D.C., 1962.
10. Dhillon, B. S. and Singh, C. *Engineering Reliability: New Techniques and Applications.* John Wiley and Sons, New York, 1981.
11. Hammer, W. "Numerical Evaluation of Accident Potentials." *Annals of Reliability and Maintainability Conference,* American Institute of Aeronautics and Astronautics, 1966, pp. 494–500.
12. Eagle, K. H. "Fault Tree and Reliability Analysis Comparison." *Proceedings of the Annual Symposium on Reliability,* Institute of Electrical and Electronics Engineers, 1969, pp. 12–17.
13. Gloss, D. S. and Wardle, M. G. *Introduction to Safety Engineering.* John Wiley & Sons, New York, 1984.
14. Salvendy, G. "Human Factors in Planning Robotic Systems," in *Handbook of Industrial Robotics.* S. Y. Nof, editor, John Wiley and Sons, New York, 1985, pp. 639–664.
15. Nertnery, R. J. and Bullock, M. G. *Human Factors in Design,* Report No. ERDA-76-45-2. The Energy Research and Development Administration, Department of Energy, Washington, D.C., 1976.
16. Dhillon, B. S. *Engineering Design: A Modern Approach.* Richard D. Irwin, Inc., Chicago, 1996.

Cost Considerations

INTRODUCTION

It is often said that competition is the best policeman of the free market. Certainly competition is a prime factor in the current trend toward ensuring that equipment be cost-effective to support across its entire life span. In many cases, the cost of acquiring a product is less than the cost of ownership over the product's life cycle. The hidden costs associated with equipment operation and support can account for as much as 75% of the total life cycle cost [1]. In some cases, support costs may, in the end, even total 10 to 100 times the original procurement cost. Cost of ownership includes operation costs (such as the cost of personnel, facilities, and utilities), maintenance costs, the cost of test and support equipment, retirement and disposal costs, technical data costs, the cost of training operations and maintenance personnel, and the cost of spares, inventory, and other support materials.

Clearly, reducing the cost of ownership is critical if equipment is to be cost-effective. The opportunity for creating savings in a product's life cycle cost decreases dramatically in the progress from the concept design and advance planning phase to the production and construction phase. In fact, 60% to 70% of the projected life cycle cost can sometimes be locked in by the completion of the preliminary design phase. This means the greatest impact on costs comes from decisions made during the early design phases.

COSTS ASSOCIATED WITH MAINTAINABILITY

Maintainability is an important factor in the total cost of equipment. An increase in maintainability can lead to reduction in operation and support costs. For example, a more maintainable product lowers maintenance time and operating costs. Furthermore, more efficient maintenance means a faster return to operation or service, decreasing downtime. Figure 8-1 shows various ways of increasing maintainability [2].

There are many components of investment cost related to maintainability. These include, as shown in Figure 8-2, the costs of prime equipment, system engineering management, repair parts, support equipment, data, training, system test and evaluation, and new operational facilities [2].

RELIABILITY COST

The following sections present some ways in which reliability costs significantly influence maintainability costs [3].

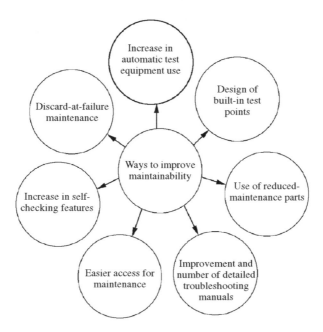

Figure 8-1. Ways to improve equipment maintainability.

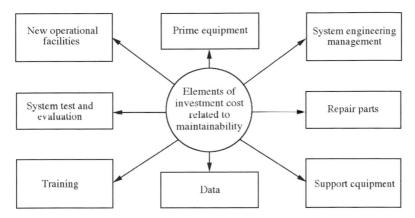

Figure 8-2. Elements of investment cost related to maintainability.

Reliability as a Capital Investment

It is possible to analyze the attractiveness that product reliability, because of the impact of reliability on price and its relationship to value, will have as a capital investment. Return on investment (ROI) is the basic measure of this attractiveness. It should be noted that ROI provides only an approximation because it assumes that product has an indefinite life and it does not take into consideration the time value of money. Nonetheless, the ROI approach can provide a "quick and dirty" assessment of the cost of reliability programs that will prevent or reduce expensive equipment failures.

Cost of Reliability from the Manufacturer's Perspective

The total cost related to reliability that a manufacturer incurs during the design, manufacture, and warranty period of a product may be called the cost of reliability. The concept of cost of reliability is related to the notion of cost of quality, whose principles were established in the 1950s and have been verified and validated since in virtually all manufacturing sectors. The cost of quality approach involves three steps: measure the economic state of quality, identify areas for quality improvement, and verify and document the effectiveness and impact of quality improvement initiatives. The elements that make up the total cost of quality—external cost of quality, internal cost of quality, appraisal cost, and defect prevention cost—easily apply to the cost of reliability, as shown in Figure 8-3: internal cost of failure, external cost of failure, failure prevention cost, and reliability appraisal cost. The internal cost of failure includes the cost of redesigning, reworking, and retesting, and the cost of equipment downtime and yield losses. Some of the components of the external cost of failure are the cost of failure analysis and of spare parts inventory, and unreliability costs during the warranty period. Failure prevention costs include reliability screening, design reviews, product qualification testing, reliability training, the development of reliability standards and guidelines, customer requirements research, and the performance of failure modes and effect analysis and fault-tree analysis. Reliability appraisal cost includes elements such as the costs of reliability modeling, life testing, abuse testing, environmental ruggedness evaluation, and failure data reporting and analysis.

The understanding of these classifications is essential for the success of reliability programs and for planning reliability assurance resources, warranty policies, training programs, and environmental and life-testing facilities. Management must not overlook the fact that cost of reliability and cost of quality are dependent variables, reflecting successes and failures of the reliability and quality programs. From the manufacturer's perspective, the total cost of reliability is expressed by

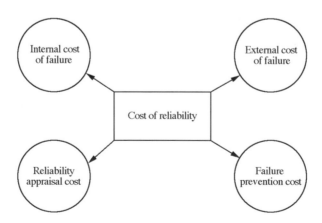

Figure 8-3. Elements of the cost of reliability.

$$\text{TCOR} = C_{rd} + C_{rm} + C_w \qquad (8.1)$$

where TCOR is the total cost of reliability.
C_{rd} is the cost of reliability design.
C_{rm} is the cost of reliability manufacturing.
C_w is the warranty cost.

DISCOUNTING FORMULAS

These formulas factor the element of time into calculations having to do with money. A sum of money received or spent today has a different buying power than the same sum received or spent months or years later. Some cost analyses therefore require this factor to be taken into consideration. For example, the maintenance cost a piece of equipment will generate two years from today may need to be discounted to its present worth. This section presents a number of discounting formulas.

Single-Payment Compound Amount

In this case, periodically or annually earned interest is added to the principal or the original amount, and interest starts accruing on the new total. Thus, the amount after one year is

$$\text{AM}_1 = \text{OM} + \text{OM}(i) \qquad (8.2)$$

where AM_1 is the amount after one year.
OM is the original amount or principal.
i is the interest rate per year.

The amount at the end of a two-year period is expressed by

$$\text{AM}_2 = \text{AM}_1 + \text{AM}_1(i) \qquad (8.3)$$

Substituting Equation 8.2 into Equation 8.3 yields

$$\text{AM}_2 = (1 + i)[\text{OM} + \text{OM}(i)] = \text{OM}(1 + i)^2 \qquad (8.4)$$

Similarly, the amount at the end of a three year-period is

$$AM_3 = AM_2 + AM_2(i) \qquad (8.5)$$

Inserting Equation 8.4 into Equation 8.5, we get

$$AM_3 = OM(1 + i)^3 \qquad (8.6)$$

Thus, the amount at the end of a period of n years is

$$AM_n = OM(1 + i)^n \qquad (8.7)$$

where n is the number of years.

This same formula also gives the future total, after inflation or other annual rates of increase, of a present amount.

Example 8-1

Assume that the estimated maintenance cost for an aircraft engine is \$400,000 in present-day dollars. Calculate the amount of this cost after five years, if the rate of increase in cost is 5% per year.
Inserting the above given data into Equation 8.7, we get

$$AM_5 = (400,000)(1 + 0.05)^5 = \$510,512.62$$

Thus, the maintenance cost for the aircraft engine after five years will be \$510,512.62.

Compound Amount of an Equal-Payment Series

This formula estimates the compound amount when a series of equal payments, also known as an annuity, are obtained or deposited at the end of each year. The compound amount is expressed by [4]

$$AM_{cn} = P[\{(1 + i)^n - 1\}/i] \qquad (8.8)$$

where AM_{cn} is the compound amount after n years.
  P is the payment made each year.
  i is the interest rate per year.

Example 8-2

Assume that the annual repair cost of a motor vehicle is \$10,000. Determine the total amount of repair costs after a 10-year period, if the rate of increase in cost is 7% per annum. Substituting the given data into Equation 8.8 yields

$$AM_{c10} = (10,000)[\{(1 + 0.07)^{10} - 1\}/0.07] = \$138,164.48$$

Thus, the total amount of the repair costs after 10 years will be \$138,164.48.

Present Value of a Single Payment

The present value of single payment from Equation 8.7 is

$$PV = OM = FW/(1 + i)^n \tag{8.9}$$

where   PV is the present value.
FW = AM_n is the future worth.

Example 8-3

It is estimated that eight years from now, the annual maintenance cost of a military tank will be \$1.2 million. Calculate the present worth of that cost, if the rate of increase in cost is 7% per annum.
Substituting the specified values into Equation 8.9 yields

$$PV = (1.2)/(1 + 0.07)^8 = \$0.6984 \text{ million}$$

The present worth of the maintenance cost is \$698,400.

Present Value of Uniform Periodic Payments

The present value of the sum of uniform payments made at the end of each of n years, that is, of the compound amount of an equal-payment series, is expressed by [4]

$$PV_a = P\left[\frac{1 - (1 + i)^{-n}}{i}\right] \tag{8.10}$$

where PV_a is the present value of the sum of uniform payments.

Example 8-4

In Example 8-2, determine the present value of the sum of annual repair costs and compare the end results of both calculations. Inserting the specified data into Equation 8.10, we get

$$PV_a = (10,000)[1 - (1 + 0.07)^{-10}] = \$70,235.80$$

Thus, the present value of the sum of annual repair costs associated with the motor vehicle will be $70,235.80, compared to the total amount of $138,164.48 after 10 years. It means that this future amount will be roughly two times the present value.

LIFE CYCLE COSTING

Due to reasons including market pressure, life cycle costing is now often used in the procurement of expensive systems or equipment. The term "life cycle costing" first appeared in a document [5] prepared for the United States Department of Defense in 1965. Life cycle cost is the sum of all costs incurred during the life time of an item, that is, the total of procurement and ownership costs. Life cycle cost analysis examines the effect on cost of alternative equipment designs. Life cycle costing plays an important role in maintainability analysis, particularly with respect to operation and maintenance costs.

Reference 4 gives a comprehensive bibliography on life cycle costing.

Life Cycle Costing Steps

The steps involved in life cycle costing are [6]:

- Estimate the useful life of the equipment.
- Estimate all associated costs, including the costs of operation and maintenance.
- Estimate the terminal value of the equipment.
- Subtract the terminal value from the ownership cost of the equipment.
- Take the result of this calculation, and find its present value.

- Obtain the life cycle cost of the equipment by adding the procurement cost to this present value amount.
- Repeat these steps for each product being considered for acquisition.
- Compare the life cycle costs of these products.
- Choose the product with the lowest life cycle cost, in balance with other considerations.

Data Sources for Life Cycle Cost Analysis and Important Points Associated with Life Cycle Costing

Sources of data for life cycle cost analysis include engineering design data, reliability and maintainability data, logistics support data, market analysis data, management planning data, production or construction data, accounting data, data related to value analysis, and consumer utilization data [1].

Some important points about life cycle costing are [4]:

- Risk management is the essence of life cycle costing.
- Management plays a key role in the success of the life cycle costing effort.
- The main aim of the life cycle costing effort is to produce maximum benefits from minimum resources.
- Good data are essential for good life cycle cost calculations.
- If shortcomings arise in the data, the superior knowledge and experience of the cost analyst may compensate for them.
- Expect surprises, irrespective of the cost analyst's abilities.
- Both the manufacturer and the user must take effective steps to control life cycle costs.
- Throughout the lifespan of the program, trade-offs must be made between life cycle cost, performance, and design to cost.
- The life cycle cost estimation model must take into consideration all relevant costs associated with the program.

Advantages and Disadvantages of Life Cycle Costing

The advantages of life cycle costing are that it is an excellent tool for comparing the cost of competing projects, controlling program costs, selecting among competing contractors, making decisions associated with equipment replacement, reducing total cost, and conducting planning and budgeting. Some of the disadvantages of life cycle costing are that it is time consuming and expensive, that collecting the data needed for analysis can be a trying task, and that the data available is sometimes of doubtful accuracy.

Life Cycle Cost Models

There are many life cycle cost estimation models available in published literature, and they fall into two groups: general models and specific models. As the name suggests, the general models can be applied to a variety of situations, while the specific models are tailored for specific applications. The data to be input into a life cycle cost model include the purchase price of the product, mean time between failures (MTBF), mean time to repair (MTTR), average material cost of a failure, labor cost per preventive maintenance action, labor cost per corrective maintenance action, installation costs, training costs, the warranty coverage period cost of carrying spares in inventory, and shipment forecasts over the course of the product's useful life [7].

General Life Cycle Cost Estimation Models

There are many general life cycle cost estimation models available in the published literature. Reference 4 presents nine such models. All these models determine the total cost of an item over its life span, but they vary in the methods they use to estimate many of the major costs used in the calculation. The following two life cycle cost models demonstrate this point.

• Life Cycle Cost Model I

This life cycle cost model calculates two major kinds of costs: recurring and nonrecurring. The life cycle cost is expressed by [8]

$$LCC = RC + NRC \tag{8.11}$$

where LCC is the product's life cycle cost.
     RC is the recurring cost.
     NRC is the nonrecurring cost.

The major elements of the recurring cost are operating cost, maintenance cost, support cost, manpower cost, and inventory cost. For the nonrecurring cost, they are procurement cost, reliability and maintainability improvement cost, research and development cost, installation cost, training cost, support cost, qualification approval cost, life cycle management cost, test equipment cost, and transportation cost. The present value of the recurring cost must be obtained, using discounting formulas, before it is added to the nonrecurring cost.

• Life Cycle Cost Model II

Three major costs form the life cycle cost in this model: research and development cost, investment cost, and operations and maintenance cost [1, 9–11]. Thus, the life cycle cost is given by

$$LCC = RDC + IC + OMC \qquad (8.12)$$

where RDC is the research and development cost.

　　　IC is the investment cost.

　　OMC is the operations and maintenance cost.

The components of the research and development cost are engineering design cost, covering system engineering, reliability, maintainability, human factors, producibility, electrical design, mechanical design, and logistic support analysis; advanced research and development cost; engineering development and test cost, for example, the cost of engineering models and of testing and evaluation; engineering data cost; and program management cost.

The investment cost consists of construction cost, that is the cost of manufacturing facilities, test facilities, operational facilities, and maintenance facilities; manufacturing cost, including the cost of manufacturing engineering, quality control, fabrication, assembly, tools and test equipment, test and inspection, materials, and packing and shipping; and initial logistic support cost, or the cost of program management, test and support equipment, initial spare and repair parts, provisioning, initial inventory, first destination transportation, technical data preparation, and initial training and training equipment.

The elements of the operations and maintenance cost are modification cost; disposal cost; operations cost, that is, the cost of operations personnel, operational facilities, support and handling equipment, and operator training; and maintenance cost, including the cost of maintenance personnel, spare and repair parts, maintenance facilities, maintenance training, maintenance of test and support equipment, transportation and handling, and technical data.

Specific Life Cycle Cost Estimation Models

As mentioned earlier, these models are tailored to meet specific needs. Reference 4 presents many examples of models. Two of them are as follows:

• Specific Life Cycle Cost Estimation Model I

This model estimates the life cycle cost of switching power supplies [12]. This cost is expressed by

$$LCC_{sp} = IC + FC \tag{8.13}$$

where LCC_{sp} is the life cycle cost of switching power supplies.
FC is the failure cost.
IC is the initial cost.

The failure cost, FC, is expressed by

$$FC = \lambda T(RC + SC) \tag{8.14}$$

where RC is the cost of repairs.
SC is the cost of spares.
λ is the constant failure rate of switching power supplies.
T is the expected life of the switching power supply.

The cost of spares SC, is given by

$$SC = (USC)(\theta) \tag{8.15}$$

where θ is the fractional quantity of spares for each active unit.
USC is the unit spare cost.

• Specific Life Cycle Cost Estimation Model II

This model was developed to estimate the life cycle cost of an early warning radar system [13]. The radar system's life cycle cost is expressed by

$$LCC_{ER} = C_a + C_O + C_1$$

where LCC_{ER} is the life cycle cost of the early warning radar system.
C_a is the acquisition cost of the system.
C_O is the operational cost of the system.
C_1 is the logistic support cost of the system.

Past experience indicates that C_a accounts for 28% of LCC_{ER}, C_O for 12% of LCC_{ER}, and C_1 for 60% of LCC_{ER}.

For the procurement cost, the breakdown percentages are 20.16% for the fabrication cost, 3.92% for installation and checkout costs, 3.36% for the design cost, and 0.56% for the documentation cost. The components of the operations cost are cost of personnel (8.04%), cost of power (3.84%), and cost of fuel (0.048%). The following breakdown percentages form the logistic support cost:

- 38.64% (repair labor cost)
- 11.04% (replacement spares cost)
- 5.52% (repair material cost)
- 3.25% (initial spares cost)
- 1.18% (age cost)
- 0.365% (cost of initial training)

Example 8-5

Assume that a company is considering buying an electric generator. Manufacturers A, B, and C are bidding to sell the system. Table 8.1 presents data for generators produced by all three manufacturers. Determine which of the three electric generators has the lowest life cycle cost.

Engineering Maintainability

Table 8.1
Life Cycle Cost Data for Electric Generators

Item	Manufacturer A's Generator	Manufacturer B's Generator	Manufacturer C's Generator
Procurement cost	$1.5 million	$2.0 million	$1.8 million
Expected useful life	15 years	15 years	15 years
Expected yearly operating cost	$80,000	$30,000	$40,000
Expected cost of a failure	$10,000	$12,000	$11,000
Failure rate	0.07 failures per year	0.08 failures per year	0.075 failures per year
Disposal cost	$30,000	$40,000	$35,000
Annual rate of increase in costs	5%	5%	5%

Manufacturer A's Electric Generator

The annual expected failure cost is

$$EFC_A = (10,000)(0.07) = \$700$$

Inserting the data from Table 8.1 into Equation 8.10, we obtain the following present value of the failure cost:

$$FC_A = (700)\left[\frac{1 - (1 + 0.05)^{-15}}{0.05}\right] = \$7,265.7$$

By inserting the data into Equation 8.9, we set the following present value of the disposal cost:

$$DC_A = PV(30,000)/(1 + 0.05)^{15} = \$14,430.5$$

Substituting the data into Equation 8.10, the present value of the operating cost is

$$OC_A = PV(80,000)\left[\frac{1-(1+0.05)^{-15}}{0.05}\right] = \$830,372.6$$

Adding these three costs to the procurement cost, the life cycle cost of the electric generator from manufacturer A is

$$LCC_A = 1,500,000 + 14,430.5 + 830,372.6 + 7,265.7 = \$2,352,068.8$$

Manufacturer B's Electric Generator

The annual expected failure cost is

$$EFC_B = (12,000)0.08 = \$960$$

Substituting the given data into Equation 8.10, we get the following present value of the failure cost:

$$FC_B = (960)\left[\frac{1-(1+0.05)^{-15}}{0.05}\right] = \$9,964.4$$

Inserting the data into Equation 8.9, we obtain the following present value of the disposal cost:

$$DC_B = PV = (40,000)/(1+0.05)^{15} = \$19,240.6$$

By substituting the data into Equation 8.10, we obtain the present value of the operating cost:

$$OC_B = PV_a = (30,000)\left[\frac{1-(1+0.05)^{-15}}{0.05}\right] = \$311,389.7$$

Adding these three costs to the procurement cost, the life cycle cost of the electric generator from manufacturer B is

$$LCC_B = 2,000,000 + 19,240.6 + 311,389.7 + 9,964.4$$
$$= \$2,340,594.74$$

Manufacturer C's Electric Generator

The annual expected failure cost is

$$EFC_C = (11,000)(0.075) = \$825$$

Substituting the given data into Equation 8.10, we obtain the following present value of the failure cost:

$$FC_C = (825)\left[\frac{1 - (1 + 0.05)^{-15}}{0.05}\right] = \$8,563.2$$

By inserting the data into Equation 8.9, we get the following present value of the disposal cost:

$$DC_C = PV = (35,000)/(1 + 0.05)^{15} = \$16,835.5$$

Substituting the given values into Equation 8.10 yields the following present value of the operating cost:

$$OC_C = PV_a = (40,000)\left[\frac{1 - (1 + 0.05)^{-15}}{0.05}\right] = \$415,186.3$$

Adding these costs to the acquisition cost, the life cycle cost of the electric generator from manufacturer C is

$$LCC_C = 1,800,000 + 16,835.5 + 415,186.3 + 8,563.2 = \$2,240,585$$

This examination of the life cycle costs of generators from companies A, B, and C shows that the one with the lowest life cycle cost comes from manufacturer C.

MAINTENANCE COST ESTIMATION MODELS

This section presents a number of models for estimating costs related to maintenance and maintainability.

Corrective Maintenance Cost Estimation Model

This model estimates the corrective maintenance labor cost for a piece of equipment. The annual cost is expressed by

$$C_{CM} = \frac{(SOH)(LC)(MTTR)}{MTBF} \qquad (8.17)$$

where SOH represents the scheduled operating hours of the equipment.
LC is the maintenance labor cost per hour.
MTBF is the mean time between failures for the equipment.
MTTR is the mean time to repair for the equipment.

Example 8-6

A heavy-duty motor is scheduled to operate for 3,000 hours annually. The expected MTBF and MTTR of the motor are 1,000 hours and 10 hours, respectively. Determine the annual labor cost of corrective maintenance for the motor, if the maintenance labor rate is $25 per hour. Substituting the given data into Equation 8.17 yields

$$C_{CM} = \frac{(3000)(25)(10)}{1000} = \$750$$

It means the yearly labor cost is $750.

Software Maintenance Cost Estimation Model

This formula estimates software maintenance costs as [14]

$$C_{SM} = \frac{3nC}{\alpha} \qquad (8.18)$$

where C_{SM} is the software maintenance cost.
α is the difficulty constant; $\alpha = 100$ (for hard programs), $\alpha = 500$ (for easy programs), and $\alpha = 250$ (for programs of medium difficulty).

n is the number of instructions to be changed per month.
C is the labor cost per man-month.

Equipment Operations and Maintenance Cost Estimation Model

This model gives an estimation of operations and maintenance cost for an equipment life cycle, expressed by

$$C_{om} = C_{lo} + C_d + C_{lm} + C_{lmn}$$

where C_{om} is the equipment operations and maintenance cost.
  C_{lo} is the cost of equipment life cycle operations.
  C_d is the equipment disposal or phase-out cost.
  C_{lm} is the cost of equipment life cycle modifications.
  C_{lmn} is the cost of equipment life cycle maintenance.

Equipment Initial Logistic Support Cost Estimation Model

The initial logistic support cost is given by

$$C_{is} = C_{ith} + C_{pm} + C_{it} + C_p + C_{td} + C_{ism} + C_{pot} + C_{ii} \tag{8.20}$$

where C_{is} is the initial logistic support cost for the equipment.
  C_{ith} is the cost of initial transportation and handling.
  C_{pm} is the logistic program management cost.
  C_{it} is the cost of initial training and training equipment.
  C_p is the provisioning cost, including preparation of procurement data for spares, test, and support equipment.
  C_{td} is the cost of technical data preparation.
  C_{ism} is the initial spare and repair parts cost.
  C_{pot} is the operational test and support equipment procurement cost.
  C_{ii} is the cost of initial inventory management.

Spare and Repair Parts Cost Estimation Model

This model estimates the cost of spare and repair parts as

$$C_{s/r} = C_c + C_{osr} + C_{ssr} + C_{dsr} + C_{isr} \tag{8.21}$$

where $C_{s/r}$ is the cost of spare and repair parts.

C_c is the cost of consumables.

C_{osr} is the organizational spare and repair parts cost.

C_{ssr} is the supplier spare and repair parts cost.

C_{dsr} is the depot spare and repair parts cost.

C_{isr} is the intermediate spare and repair parts cost.

Equipment Maintenance Cost Estimation Model

This model calculates the cost of equipment maintenance with the formula [15]

$$MC = PMC + CMC + SPIC \tag{8.22}$$

where MC is the equipment maintenance cost.

PMC is the cost of preventive maintenance.

CMC is the cost of corrective maintenance.

SPIC is the cost of spare parts inventory.

The cost of preventive maintenance, PMC, is defined by

$$PMC = \frac{(ST_{pm} + TT_{pm})(UH)R}{SI_{pm}} \tag{8.23}$$

where ST_{pm} is the scheduled time preventive maintenance work will take.

TT_{pm} is the expected travel time for preventive maintenance.

SI_{pm} is the scheduled interval at which preventive maintenance takes place.

UH is the number of usage hours, or in-use time, per time period considered.

R is the servicing engineer's hourly rate, including the prorated parts cost.

Similarly, the corrective maintenance cost, CMC, is expressed by

$$CMC = \frac{(TT_{cm} + MTTR)(UH)R}{MTBF} \qquad (8.24)$$

where TT_{cm} is the expected travel time for corrective maintenance.
MTTR is the mean time to repair for the equipment.
MTBF is the mean time between failures for the equipment.

The cost of spare parts inventory, SPIC, is given by

$$SPIC = (OMC)(ICR) \qquad (8.25)$$

where OMC is the original manufacturing cost of spare parts.
ICR is the inventory rate, expressed as a percentage, including such factors as interest, handling cost, and depreciation, etc.

Example 8-7

Assume that for maintenance of a personal computer, the following values are given:

- MTTR = 2 hours
- MTBF = 7,500 hours
- UH = 4,500 hours per annum
- R = $400 per hour
- ICR = 8% per year
- OMC = $1,000
- SI_{pm} = 2,500 hours
- SI_{pm} = .35 hour
- TT_{pm} = 0.25 hour
- TT_{cm} = 0.25 hour

Determine the annual maintenance cost for the personal computer. Substituting the given values into Equation 8.25 yields

$$SPIC = (1,000)(0.08) = \$80$$

Using Equation 8.23, we get

$$PMC = \frac{(0.35 + 0.25)(4,500)(400)}{2,500} = \$432$$

Equation 8.24 yields

$$CMC = \frac{(0.25 + 2)(4,500)(400)}{7,500} = \$540$$

Inserting these three values into Equation 8.22 yields

$$MC = 432 + 540 + 80 = \$1,052$$

The annual maintenance cost is $1,052.

MAINTAINABILITY, MAINTENANCE COSTS, AND COST COMPARISONS

The level of maintainability of a product determines the kinds of maintenance work that can and will need to be performed at each point in the product's life cycle, and the difficulty and expense of performing them. Maintainability features, such as mean time to repair (MTTR), therefore influence maintenance costs such as required manpower. For example, if the design calls for the inclusion of built-in test equipment, the time to fault detection and isolation should be lower. Usually, higher maintainability means less required maintenance, and therefore lower maintenance costs. In early equipment design, several alternative levels of built-in test equipment and other factors that can reduce maintenance costs should be considered.

The objective of performing an economic trade-off analysis is to determine all costs for each alternative under consideration and then to compare them. Usually, the alternative with the lowest cost should be selected. This approach is also useful in determining whether items should be designed to be thrown away or to be repaired. The factors include the cost of hardware, manpower, training, test equipment and tools, and repair facilities, replacement parts, packaging and shipping, repair parts, and supply, administration, and cataloging [2, 16].

PROBLEMS

1. What is the difference between reliability cost and maintainability cost?
2. Discuss in detail the elements that make up the cost of reliability.
3. Five years from now the annual maintenance cost for an electrical transformer is estimated to be $250,000. Determine the present value of this cost, if the rate of increase in cost is expected to be 8% per year.
4. Describe the term "life cycle costing" and the steps associated with life cycle costing.
5. What are some sources for the data used in life cycle cost analysis?
6. Discuss the advantages and disadvantages of life cycle costing.
7. What is the difference between general life cycle cost models and specific life cycle cost models?
8. Discuss the impact of equipment maintainability cost on life cycle cost.
9. A pulverizer is scheduled to operate for 4,000 hours per year and its associated MTBF and MTTR are 1,200 hours and 5 hours, respectively. Calculate the annual labor cost of corrective maintenance for the pulverizer, if the estimated maintenance labor rate is $30 per hour.
10. An organization is planning to buy a motor vehicle and has received bids from two manufacturers. The costs and other data related to the motor vehicles are given in Table 8.2. Determine which of the motor vehicles the organization should acquire from the point of view of life cycle cost.

Table 8.2
Cost and Other Related Data for the Two Motor Vehicles

Item	Manufacturer A's Motor Vehicle	Manufacturer B's Motor Vehicle
Acquisition cost	$80,000	$60,000
Annual operating cost	$5,000	$4,000
Expected cost of a failure	$1,000	$900
Expected useful life	12 years	12 years
Annual failure rate	0.05 failures per year	0.055 failures per year
Annual rate of increase in costs	7%	7%
Disposal cost	$600	$500

REFERENCES

1. Blanchard, B. S., Verma, D. and Peterson, E. L. *Maintainability: A Key to Effective Serviceability and Maintenance Management.* John Wiley and Sons, New York, 1995.
2. AMCP 706-133, *Engineering Design Handbook: Maintainability Engineering Theory and Practice.* Department of Defense, Washington, D.C., 1976.
3. Kohoutek, H. "Economics of Reliability," in *Handbook of Reliability Engineering, and Management.* W. Grant Ireson, C. F. Coombs, and R. Y. Moss, editors, McGraw-Hill, New York, 1996, pp. 4.1–4.26.
4. Dhillon, B. S. *Life Cycle Costing: Techniques, Models and Applications.* Gordon and Breach Science Publishers, New York, 1989.
5. LMI Task 4C-5, *Life Cycle Costing in Equipment Procurement.* Logistics Management Institute, Washington, D.C., 1965.
6. Coe, C. K. "Life Cycle Costing by State Governments." *Public Administration Review,* September/October, 1981, pp. 564–569.
7. Seiwiorek, D. P. and Swarz, R. S. *The Theory and Procedures of Reliable System Design.* Digital Press, Digital Equipment Corporation, Bedford, Massachussets, 1982.
8. Reiche, H. "Life Cycle Cost," in *Reliability and Maintainability of Electronic Systems.* J. E. Arsenault and I. A. Roberts, editors; Computer Science Press, Potomac, Maryland, 1980, pp. 3–23.
9. Pamphlet No. 11-2, *Research and Development Cost Guide for Army Material Systems.* Department of Defense, Washington, D.C., 1976.
10. Pamphlet No. 11-3, *Investment Cost Guide for Army Material Systems.* Department of Defense, Washington, D.C., 1976.
11. Pamphlet No. 11-4, *Operating and Support Cost Guide for Army Material Systems.* Department of Defense, Washington, D.C., 1976.
12. Monteith, D. and Shaw, B. "Improved R, M, and LCC for Switching Power Supplies." *Proceedings of the Annual Reliability and Maintainability Symposium,* 1979, pp. 262–265.
13. Eddins-Earles, M. *Factors, Formulas, and Structures for Life Cycle Costing.* Eddins-Earles, Concord, Massachusetts, 1981.
14. Sheldon, M. R. *Life Cycle Costing: A Better Method of Government Procurement.* Westview Press, Boulder, Colorado, 1979.
15. Grant Ireson, W., Coombs, C. F. and Moss, R. Y. *Handbook of Reliability Engineering and Management.* McGraw-Hill Book Company, New York, 1996.
16. Gordman, A. S. and Slattery, T. B. *Maintainability: A Major Element of System Effectiveness.* John Wiley and Sons, New York, 1964.

Reliability-Centered Maintenance

INTRODUCTION

Reliability-centered maintenance (RCM) systematically identifies the preventive maintenance tasks required to sustain, in the most cost-effective manner possible, the maximum level of reliability and safety that can be expected from a product when it receives effective maintenance.

The history of RCM began within the commercial aircraft industry in the late 1960s. A 1968 handbook titled "Maintenance Evaluation and Program Development," prepared by the United States Air Transport Association (ATA) for use with the Boeing 747 aircraft, contained one of the earliest formal treatments of the subject [1–3]. Two years later, a revised version of the handbook also discussed two other wide-body aircraft, the DC-10 and L-1011 [4]. It was in 1974, when the United States Department of Defense commissioned United Airlines to prepare a document on civil aviation aircraft maintenance programs, that the term "reliability-centered maintenance" was coined as the title of the resulting document [5].

In 1980, the ATA revised its second handbook to include maintenance programs for the Boeing 756 and 767 aircraft [6]. The document had a European counterpart that also covered the A-300 and Concorde aircraft [3].

In the early 1970s, after RCM methodology attracted the attention of the United States armed forces, the Navy applied the concept to

its S-3, P-3, and F-4J aircraft. Two important military documents concerning RCM appeared in 1985 [7, 8], and in 1983, the Electric Power Research Institute recommended the evaluation of RCM for applications in nuclear power plants. Today, RCM is used in countries around the world, including the United Kingdom, Canada, the United States, Spain, Singapore, and Australia.

THE DEFINITION OF RELIABILITY-CENTERED MAINTENANCE

Reliability-centered maintenance determines the maintenance needs of any facility, system, or equipment in its operating context [5]. The process entails asking questions on the following subjects:

- The functions and related performance standards of the asset in its current operating context
- Possible ways in which the asset may fail to perform its required functions
- Causes of each functional failure
- Events that follow each failure
- Significance of each failure
- Measures to prevent failure
- Corrective measures that may be taken if there is no appropriate preventive step

THE RCM PROCESS

The RCM process takes place first during the equipment design and development phase, when it is used to develop maintenance plans.

During product operation and deployment, these plans are then modified based on field experience. The following two criteria are key to the maintenance plans [9]:

- **Parts that are not critical to safety.** In this case, preventive maintenance tasks should be chosen that will decrease the owner-ship life cycle cost.
- **Parts that are critical to safety.** In this case, preventive main-tenance tasks should be chosen that will help prevent reliability

or safety from dropping to an unacceptable level, or will help reduce the ownership life cycle cost.

It is through the preventive maintenance program that incipient failures are detected and corrected, the probability of failure is reduced, hidden failures are detected, and the cost-effectiveness of the maintenance program is improved.

Figure 9-1 shows the seven steps that make up the basic RCM process.

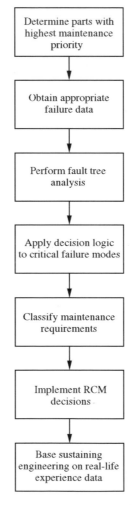

Figure 9-1. The basic steps in the RCM process.

Determine Parts with the Highest Maintenance Priority

Traditionally, failure mode and effect analysis has been employed to identify those parts whose failure would have the most significant effects. The result of this effort, in turn, provided the basis for defining the most important preventive and corrective maintenance requirements. Fault tree analysis (FTA), however, has proven more effective, because it uses the results of data collection programs and field experience to develop new, and to also upgrade existing, maintenance programs.

FTA identifies parts that are critical to safety and provides quantitative failure-mode data about them. It is also more rigorous and accurate in determining the root causes of failures and their consequences with respect to product safety. Furthermore, FTA defines the impact of intervention and control procedures in a meaningful and mathematically tractable form. Along with its role in developing and evaluating optimum maintenance requirements, FTA can also be used to identify critical parts and the appropriate preventive measures to take before accidents or incidents occur, to determine the effect of changes to the maintenance program before and after their implementation, and to perform other similar kinds of analysis.

Obtain Appropriate Failure Data

Each step in the fault tree requires data. The most important data are failure probabilities and assessments of the criticality of the failures, part failure rates, probability of operator error, and inspection efficiency data. Part failure rate data can come from experience, banks of generic failure data, and other sources.

The data banks for operator error data fall into three categories [10–11]: experimentally based data banks, field-based data banks, and subjectively based data banks. The experimentally based banks contain data gathered in the laboratory [12]. The field-based banks are based upon data gathered during operations and so often provide more realistic information. References 13 and 14 describe two such data banks. The subjectively based banks contain data generated by techniques such as DELPHI [15].

Perform Fault Tree Analysis

At this step, the analyst computes the probabilities that fault events—basic, intermediate, and top events—will occur, using the combinatorial properties of the logic elements in the fault tree. Sensitivities are calculated by assigning a probability of unity to a basic or elementary fault and then determining resultant probability of a safety incident. These sensitivities can then be used to compute each basic fault's criticality.

Criticality measures the relative seriousness or impact each fault would have on the top event of the fault tree. It also involves qualitative and quantitative analyses of the fault tree and provides a basis for ranking faults in order of their severity.

Quantitatively, criticality is [16]:

$$K = P(x)P(F|x) \tag{9.1}$$

where K is the criticality of the fault.
   P(x) is the probability that the fault will occur.
   P(F|x) is the sensitivity or conditional probability, that is, the probability that the occurrence of a fault will lead to a safety incident.

Apply Decision Logic to Critical Failure Modes

This step involves asking standard assessment questions and using the results to determine what the most effective preventive maintenance tasks would be. Each question requires a simple "yes" or "no" answer, which is then recorded on a worksheet or a computer database. After the fault tree analysis identifies the critical failure modes, decision logic is used to assess the relationship between each failure mode and each part with a high maintenance priority. The next step is to establish what maintenance tasks are necessary to prevent or reduce the incidence of each failure mode. The tasks considered necessary, and the appropriate intervals at which they should be performed, make up the overall scheduled preventive maintenance program. The decision logic used in this process consists of two levels:

- **Level 1:** This level of analysis assigns each failure mode to a category: evident threat to safety; hidden threat to safety; potential operational and economic problem; and problem with economic, but no safety or operational, consequences. Four questions should be asked: Does the failure, or a resulting failure, cause a safety incident? Can operators detect the faults or failures? Does the fault or failure lead to a direct adverse effect on operating performance? Does the hidden failure alone, or in combination with an additional failure of a system-related or backup function, lead to a safety incident?
- **Level 2:** This level uses causes for each failure mode to select needed maintenance tasks. Some examples of questions asked at this stage are: Is there a relevant and effective repair task that would lower the failure rate? Is there a relevant and effective servicing task? Is there a relevant and effective operator monitoring task? Is there a relevant and effective combination of tasks?

Classify Maintenance Requirements

This step uses the decision logic of the preceding step to sort the preventive maintenance requirements into three classifications and define a maintenance task profile. The three classifications are:

- **Hard-time maintenance requirements.** These are scheduled removals or replacements of equipment or parts at predetermined intervals of age or usage.
- **Condition-monitoring maintenance requirements.** These are unscheduled tests or inspections conducted on parts when failure of the parts can be tolerated during equipment operation or where impending failure can be discovered through routine monitoring during usual operations.
- **On-condition maintenance requirements.** These are scheduled inspections or tests that measure part deterioration. The level of part deterioration determines whether corrective maintenance should be performed or whether the part should remain in service.

The maintenance task profile contains preventive maintenance tasks selected from the Level 1 and Level 2 decision logic questions for which the answer was "yes." These tasks are organized by part number and failure mode. The profile determines the preventive maintenance tasks to be performed on each part in question.

Implement RCM Decisions

This step is concerned with setting and enacting the maintenance tasks and their frequencies. Because of the critical importance of implementation, a following section discusses it in detail.

Base Sustaining Engineering on Real-Life Experience Data

During the rest of the asset's life cycle, the RCM process focuses on reducing the burden of scheduled maintenance and cost of support while keeping the equipment in a desirable state of readiness. Once the system is operating and real life data begin to accumulate, the goal is to review previous decisions in order to eliminate excessive maintenance costs while maintaining established and desirable reliability and safety levels.

RCM IMPLEMENTATION

All the effort spent on the RCM process will become worthless if implementation is not implemented with care. The following two approaches can be used [5]:

- **Approach I:** This short-term approach focuses primarily on assets and processes and less on the people involved with them. Organizations seeking the fastest return on time and money invested in an RCM project often choose this approach, assembling appropriate specialists and asking them to concentrate only on the assets. Two options are a task force procedure and a selective procedure.

    The task force procedure assumes that the quickest and biggest returns come when RCM focuses on any assets or processes that have intractable problems with serious consequences. A small task force performs a comprehensive RCM analysis of the system, usually working full-time until the review is completed. The task force procedure can also be used simultaneously with Approach II.

The selective procedure secures a quick return, when the organization is not suffering from any acute problem, by applying the RCM process to the assets most likely to benefit from it. The procedure involves identifying "non-significant" assets that are not likely to benefit significantly from the RCM process, ranking the significant assets in order of importance, and deciding if a "template" approach—using the analysis of one asset as a "template" for the analysis of another—should be used for similar assets.

Two advantages of Approach I are that it is quick and easy to manage, because only one or two groups, consisting of a small number of people, are involved. But because this approach tends not to secure the long-term involvement and commitment of the entire organization, its end results are less likely to endure.

- **Approach II:** This long-term approach takes advantages of the opportunities RCM offers on both the human and technological fronts. The process is used to improve teamwork between the users and maintainers of the assets, to improve the performance of the assets themselves, and to improve the knowledge and motivation of the individuals involved. This approach requires a much greater commitment of resources and management time than Approach I and the cooperation and involvement of a larger number of individuals.

  This approach not only improves individual motivation and teamwork but also ensures that results will last longer. Its disadvantages are that, because a greater number of people will be involved and will need to become familiar with RCM methodology, it is slower and more difficult to manage.

RCM REVIEW GROUPS

A typical RCM review group includes a facilitator, engineering supervisor, operations supervisor, craftsman, operator, and, if required, an external specialist. It is important that each member of the review group be trained in the RCM process and have a thorough knowledge of the asset under review.

The facilitators are highly trained in RCM and the most important people in the RCM review process. They guide the work of the review groups and ensure that RCM is applied correctly and effectively, no important equipment/item is overlooked, the review meetings make real progress, and all necessary documentation is correct and completed in a timely fashion.

Soon after the review for each major piece of equipment is completed, managers with overall responsibility for the equipment must satisfy themselves that the review was performed correctly and that they support the failure consequence assessment and the selection of maintenance tasks. Managers may delegate this audit to someone with appropriate specialized knowledge.

METHODS OF MONITORING EQUIPMENT CONDITION

Because the monitoring of deviations from "normal" equipment conditions requires finer perceptions than the human senses can provide, special instruments are employed for this purpose. Condition monitoring techniques fall into one of six categories, according to the symptoms or potential failure effects they monitor [5]: dynamic effects, electrical effects, physical effects, temperature effects, particle effects, and chemical effects.

Dynamic Effects

The methods within this classification detect failures, particularly of rotating equipment, that result in abnormal energy emissions in the form of waves—for example, vibrations, pulses, or noise. Dynamic monitoring techniques include:

- **Broad band vibration analysis.** On devices such as engines, shafts, electric motors, gearboxes, and pumps, this technique monitors changes in vibration characteristics caused by problems such as wear, fatigue, mechanical looseness, or misalignment. It is an inexpensive method that requires minimal skill and data logging, and is effective in detecting simple defects. But it provides limited defect identification capability and only crude overall measurements.

- **Shock pulse monitoring.** This technique monitors surface deterioration and lack of lubrication in devices such as pneumatic impact tools, internal combustion, and rolling element bearings. The equipment required is portable and simple to use. But this method is unsuitable for slow-moving machinery with high levels of product impact noise.
- **Proximity analysis.** This method is used for devices such as fans, shafts, and motor assemblies. The problems it tracks include misalignment, rubs, and oil whirl. While it is simple and straightforward and uses portable equipment, the method's disadvantages are its limited diagnostic capability and the lengthy time it requires for analysis.
- **Real time analysis.** In devices such as gearboxes, rotating machines, and shafts, this approach monitors shock, transient, acoustic, and vibrational signals. It is capable of simultaneously analyzing bands of frequencies over the entire analysis range, and short-duration signals such as transient vibration and shocks. It also provides instantaneous continuously updated graphical displays of analyzed spectra. However, it requires expensive and nonportable equipment, high-level skills, and off-line analysis.
- **Ultrasonic leak detection.** This method detects leaks and other sources of very high frequency noise in heat exchangers, air-operated contractors on electric traction control, underground tanks, and steam condensers. The equipment is portable and can be used in highly noisy areas. But this method cannot indicate the size of a leak, and underground tanks can only be tested within a vacuum.
- **Kurtosis.** This technique monitors shock pulses in gears and rolling element bearings. It is simple to use, involves portable equipment, and is applicable to any materials with hard surfaces. But the technique can be too sensitive, limited in its applications, and significantly affected by noise from other sources.

Electrical Effects

Three techniques used to monitor electrical effects are:

- **Electrical resistance (corrometer).** In facilities such as process plants, paper mills, petroleum refineries, and gas transmission plants, the technique detects integrated metal loss, including total corrosion. The interpretation of results is normally easy, and the method is applicable in any environment and yields both total metal loss and corrosion rate data when plotted against a time scale. The main disadvantage of the technique is that it provides no indication of whether the rate of corrosion at a specific time is high or low.

- **Linear polarization resistance (corrator).** This approach, used for nuclear power heat exchange water, cooling water systems, and geothermal power generating systems, monitors the rate of corrosion in electrically conductive corrosive fluids. Among the advantages of this approach are that it is sensitive to corrosion rates as low as a fraction of a millimeter per year, provides a direct indication of corrosion rate and pitting tendency, and generates results that are easily interpretable.

- **Potential monitoring.** In materials of stainless steel, titanium, and nickel-based alloys, this method detects problems such as stress-corrosion cracking, pitting corrosion, and selective phase corrosion. It responds quickly to change and monitors localized attacks, but it does not provide a direct measure of corrosion rate or total corrosion, and it can be influenced by changes in temperature and acidity.

Physical Effects

Techniques that monitor physical effects include:

- **Magnetic particle inspection.** This technique, used for ferromagnetic materials including welds, shafts, boilers, and machined surfaces, monitors surface and near-surface cracks and discontinuities caused by wear fatigue, heat treatment, and other problems. It is a widely used, sensitive, and reliable technique, but it is also time-consuming, contaminates clean surfaces, and does not detect cracks deeper than the surface or near surface.

- **X-ray radiography.** Applied to items including compressors, welds, gearboxes, pumps, and steel structures, this technique monitors surface and subsurface discontinuities caused by gas porosity, stress, or fatigue, and also monitors discontinuities such as loose wires. While it detects defects in parts or structures hidden from view, and provides a permanent record of these problems, this technique often has a low sensitivity for crack-like defects.

- **Strain gauges.** This technique monitors strain in civil engineering structures such as tunnels and bridges. The gauge can be readily attached to almost any surface, but it must be compatible with both the material being tested and the operating environment.
- **Eddy current testing.** Applied to ferrous materials used in items like heat exchanger tubes, railway lines, boiler tubes, and hoist ropes, this method monitors factors such as material hardness and surface and subsurface discontinuities caused by wear, stress, and fatigue. It provides high defect detection sensitivity, can be used without surface preparation, and is applicable to a wide range of conducting materials, but receives poor response from non-ferrous materials.
- **Electron fractography.** This technique tracks the growth of fatigue cracks in motor vehicles, metallic components in aircraft, industrial equipment, and similar items. The failure analysis offers a high degree of certainty and the fracture surface experiences no damage when a replica is made. But the microscope equipment required is costly and its results can only be read by a specialist.
- **Ultrasonics (a pulse echo technique).** This technique is used on welds, boiler tubes, compressors, receivers, steel structures, and other items either of ferrous or non-ferrous materials. It monitors the thickness of materials subject to wear and corrosion, as well as surface and below-surface discontinuities caused by factors such as inclusions, fatigue, heat treatment, and lamination. While the technique is applicable to the majority of materials, it is difficult to differentiate the types of defects identified by it.

Temperature Effects

Three methods of monitoring temperature effects are:

- **Temperature-indicating paint.** This paint, applied to hot spots or potential points of insulation failures, reacts to surface temperature. While it is simple to use and provides a permanent record of the highest temperature reached at the point, it is only useful at two fixed temperatures and after signaling a temperature change it does not return to its original color.

- **Fiber loop thermometry.** This technique is used for items such as engines, power cables, transformer windings, and pipelines. It monitors temperature variations caused by insulation deterioration, blocked cooling systems, leaks, or other problems. This technology is operable in hazardous environments, reachable in otherwise inaccessible locations, and unaffected by the presence of electromagnetic interference. But it is also uneconomical in small installations.
- **Thermography.** This method, which is applied to items including transformers, hydraulics, building insulation, and electrical switch gears, identifies changes in heat transfer characteristics due to delamination of laminated materials or to variation in temperature caused by fatigue, leaks, wear, or other problems. The equipment is portable and quick to use, and can examine stationary or moving objects at a distance without touching them or influencing their temperature. But the equipment is costly, and the results can only be interpreted by specialists.

Particle Effects

Techniques to monitor particle effects include:

- **Magnetic chip detection.** This technique monitors wear and fatigue in equipment, such as aircraft engines, gearboxes, compressors, and turbines, with enclosed lubricating systems. It is an inexpensive method of monitoring the liquid contamination, and the debris analysis requires only a low-powered microscope. But in this case as well, a specialist is needed to interpret the results.
- **Blot testing.** This method detects fatigue, wear, and corrosion particles in circulating oil systems such as compressors, gearboxes, and engine sumps. The test is easy to set up but requires 24 hours for the oil to blot.
- **Ferrography.** This is another technique for monitoring fatigue, wear, and corrosion in enclosed lubricating and hydraulic oil systems, such as engine sumps, gearboxes, and hydraulics. Its advantages are that it measures particle shapes and sizes and is more sensitive than emission spectrometry at initial stages of

engine wear. But it measures only ferromagnetic particles, requires an electron microscope for an in-depth analysis, and is not an on-line technique.

- **X-ray fluorescence.** This also is used for enclosed lubricating and hydraulic oil systems, such as engine sumps, gearboxes, and hydraulics. It monitors wear and damage to filters, and while the instrument is capable of detecting very small traces of impurities, it is also very costly.
- **Graded filtration.** This last technique used for enclosed lubricating and hydraulic oil systems, such as hydraulics, gearboxes, and engine sumps, detects particles in lubricating oil that stem from fatigue, corrosion, and wear. This is a relatively inexpensive technique that can determine whether wear is normal or not. But its drawbacks are that it is difficult to identify particle elements, special skills are required to interpret test results, and it is not an on-line technique.

Chemical Effects

Some of the techniques available to monitor chemical effects are:

- **Gas chromatography.** This method detects gases emitted as the result of faults in nuclear power systems and turbine generators. It provides a high level of sensitivity, but drawbacks are the possibility that any fault gases in large systems will rapidly dilute and difficulty in obtaining satisfactory samples for sensitivity analysis.
- **Thin-layer activation.** This technique monitors wear in devices such as turbine blades, electrical contacts, bearings, rails, and cooling systems. Its advantage is that the wear can be measured during normal plant operation, but reactivation is needed every four years.
- **Infrared spectroscopy.** In enclosed oil systems such as compressor sumps, transformers, and engine sumps, this method measures fluid degradation and the presence of gases such as hydrogen, carbon monoxide, and methane. It offers high sensitivity and rapid analysis, but that analysis does require considerable experience and skill.
- **Spectrometric oil analysis procedure.** This technique, used with circulating oil systems, tracks wear, leaks, and corrosion. The

atomic absorption spectrometer required is relatively inexpensive but also slow and laborious to use.

RCM APPLICATIONS AND ACHIEVEMENTS

Today, RCM methodology is being applied in many sectors, including commercial aircraft production [9], the military [3, 17], and power generation [9, 18].

- **Commercial aircraft.** The benefits of practicing RCM have been clearly felt in the aircraft industry. For example, under traditional maintenance policies the initial maintenance program for the McDonnell Douglas DC-8 airplane required scheduled overhaul for a total of 339 items. However, under the RCM program, the items requiring scheduled overhaul on the DC-10 were reduced to seven. Probably the most important item affected was the DC-10 turbine propulsion engine. The elimination of scheduled overhauls for the engine created savings in labor, material costs, and a reduction of more than 50% in the spare engine inventory required. Since each engine costs more than $1 million, this translates into significant savings [9].
- **Nuclear power generation.** The Florida Power & Light Company practiced the RCM process at two of its reactors and reported that 24 actions developed through the process were not in its existing preventive maintenance program. The company estimated that these tasks could reduce preventive maintenance man-hour costs by approximately 40% and material costs by 30%. Furthermore, on the basis of the experience other industries had with RCM, the company projected additional savings of approximately 30% to 40% in corrective maintenance as well as a reduction in the rate of forced outage due to component cooling water failure [19].

Figure 9-2 shows the many benefits that application of reliability-centered maintenance methodology offers [5].

Improvement in Operating Performance

Usually, equipment performance is composed of three elements: availability, efficiency, and yield. Thus, overall plant performance is expressed by

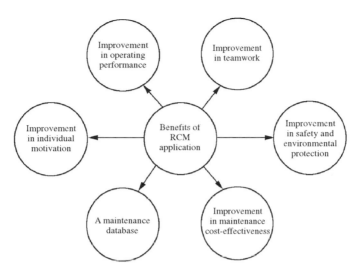

Figure 9-2. The advantages of RCM application.

$$OPP = (A)(E)(Y) \tag{9.2}$$

where OPP is the overall plant performance.
     A is availability.
     E is efficiency
     Y is yield.

The RCM process helps improve plant performance in the following ways:

- The emphasis on on-condition tasks ensures that potential failures are highlighted before they become functional failures.
- This emphasis can also reduce the frequency of major overhauls, thus improving the long-term availability of equipment.
- Eliminating superfluous facilities, equipment, or components results in a corresponding increase in reliability.
- Because it relates each failure mode to the corresponding functional failure, the information sheet becomes a tool for quick failure diagnosis. This ultimately leads to shorter repair times.
- Systematically reviewing the operational consequence of every failure that was not dealt with as a safety hazard, and employing

stringent criteria to determine task effectiveness, makes it possible to select only the most effective tasks to address each failure mode.
  • Using the people most knowledgeable about equipment to analyze failure modes helps ensure that chronic failures are identified and eliminated and that necessary preventive measures are taken.

Improvement in Maintenance Cost-Effectiveness

In many industries, maintenance forms the largest portion of operating cost after raw materials and direct production labor or energy. It has taken even first or second place in some areas. This means that controlling maintenance costs is central to cost-effectiveness. The RCM process helps lower or at least control the maintenance cost growth rate, because it reduces routine maintenance and the need for costly experts, improves purchasing of maintenance services, and provides clearer directions for acquiring new maintenance technology.

Improvement in Teamwork

Today, teamwork has become an important element in the success of many organizations. The practice of RCM not only fosters teamwork within the RCM review groups but also helps to improve communication and co-operation among various people and units: design engineers, equipment users and maintainers; production and operation units and maintenance personnel; and management, supervisors, technical personnel, and operators.

Improvement in Safety and Environmental Protection

The ways in which the practice of RCM leads to improved safety and environmental protection include:

  • The reduction in the total number and frequency of routine tasks automatically lowers the chance of critical failures occurring either during maintenance or shortly after equipment start-up.
  • The RCM decision process demands that all potential failures that would or could affect safety or the environment be eliminated or addressed.

- Examination of a failure's safety and environmental implications prior to considering its operational effects makes safety and environmental integrity a top priority.
- The attention to hidden failures and the systematic approach to failure-finding results in considerable improvement to preventive maintenance. The probability that multiple failures having serious consequences will occur is thereby substantially reduced.

Improvement in Individual Motivation

RCM helps improve the motivation of individuals involved in the review process by providing:

- A clearer understanding of an asset's functions, and of the expectations placed on each individual who works with it, helps enhance his or her competence and confidence.
- A clearer understanding of the issues beyond the control of each individual enables him or her to work more comfortably within the framework of those limitations.
- Knowledge about each group member's part in formulating goals, and in deciding what actions are required to achieve them and who should perform these actions, leads to a strong sense of ownership.

A Maintenance Database

The RCM information worksheets constitute a comprehensive maintenance database, with benefits including more accurate drawings and manuals, greater ability to adapt to changing circumstances, the introduction of expert systems, and reduced effects of manpower turnover.

REASONS FOR RCM FAILURES

The RCM methodology can generate fast results and many benefits if it is applied effectively. But not every application of RCM may yield its full potential. In fact, some may achieve very little or nothing. Some of the reasons for this may be that the application was superfluous or hurried, the analysis was conducted at too low a level, or that too much emphasis was placed on failure data, such as mean time between failures and mean time to repair [5].

PROBLEMS

1. Write an essay on the historical development of the RCM methodology.
2. Describe the steps associated with the RCM process.
3. Discuss the application of the fault tree methodology in RCM.
4. Describe two approaches associated with RCM implementation.
5. Discuss the following:
   • RCM facilitators
   • RCM review groups
6. Describe the six categories of equipment condition monitoring methods.
7. Compare the following two equipment condition monitoring techniques:
   • X-ray radiography
   • Infrared spectroscopy
8. Discuss two important areas of the RCM application.
9. List the advantages of application of the RCM methodology.
10. What are the reasons for RCM failures?

REFERENCES

1. MSGI, *Maintenance Evaluation and Program Development.* 747 Maintenance Steering Group Handbook, Air Transport Association, Washington, DC., 1968.
2. Smith, A. M., Vasudevan, R. V., Matteson, T. D. and Gaertner, J. P. "Enhancing Plant Preventive Maintenance Via RCM." *Proceedings of the Annual Reliability and Maintainability Symposium,* 1986, pp. 120–125.
3. Anderson, R. T. and Neri, L. *Reliability Centered Maintenance: Management and Engineering Methods.* Elsevier Applied Science Publishers, London, 1990.
4. MSG2, *Airline/Manufacturer Maintenance Program Planning Document.* Air Transport Association, Washington, D.C., 1970.

5. Moubray, J. *Reliability-Centered Maintenance.* Industrial Press, Inc., New York, 1992.

6. MSG3, *Airline/Manufacturer Maintenance Program Planning Document.* Air Transport Association, Washington, D.C., 1980.

7. MIL-STD-1843, *Reliability Centered Maintenance for Aircraft, Engines, and Equipment.* Department of Defense, Washington, D.C., 1985.

8. US AMC Pamphlet 750-2, *Guide to Reliability Centered Maintenance.* Department of Defense, Washington, D.C., 1985.

9. Brauer, D. C. and Brauer, G. D. "Reliability Centered Maintenance." *IEEE Transactions on Reliability,* Vol. 36, 1987, pp. 17–24.

10. Dhillon, B. S. and Singh, C. *Engineering Reliability: New Techniques and Applications.* John Wiley and Sons, New York, 1981.

11. Meister, D. "Subjective Data in Human Reliability Estimates." *Proceedings of the Annual Reliability and Maintainability Symposium,* 1978, pp. 10–15.

12. Munger, S. J., Smith, R. W. and Payne, D. *An Index of Electronic Equipment Operability: Data Store,* Report AIR-C43-1/62-RP (1). American Institute for Research, Pittsburgh, Pennsylvania, 1962.

13. Swain, A. D. "Development of a Human Error Rate Data Bank." *Proceedings of the U.S. Navy Human Reliability Workshop,* Report No. NAVSHIPS 0967-412-4010, Department of Defense, Washington, D.C., 1977.

14. Urmston, R., *Operational Recording and Evaluation Data System (OPREDS),* Descriptive Brochures, Code 3400. Navy Electronics Laboratory Center, San Diego, California, 1971.

15. Dalkey, N. and Helmer, F. "An Experimental Application of the DELPHI Method to the Use of Experts." *Management Science,* Vol. 9, 1963, pp. 458–467.

16. *Fault Tree Handbook,* Report No. NUREG-0492. U.S. Nuclear Regulatory Commission, Washington, D.C., 1981.

17. Hollick, L. H. and Nelson, G. N. "Rationalizing Scheduled-Maintenance Requirements Using Reliability Centered Maintenance—A Canadian Air Force Perspective." *Proceedings of the Annual Reliability and Maintainability Symposium,* 1995, pp. 11–17.

18. Smith, A. M. "Preventive Maintenance Impact on Plant Availability." *Proceedings of the Annual Reliability and Maintainability Symposium,* 1992, pp. 177–180.

19. "Equipment Reliability Sets Maintenance Needs." *Electrical World,* August 1985, pp. 50–51.

CHAPTER

10

Maintainability Testing, Demonstration, and Data

INTRODUCTION

The primary function of maintainability testing and demonstration is to verify the maintainability features that have been designed and built into a product [1]. Testing and demonstration also provide the customer with confidence, prior to making production commitments, that the equipment design under consideration satisfies the maintainability requirements. Prior to the testing and demonstration phase, the tasks of the maintainability program have been basically analytical. They have provided a certain degree of assurance—through steps such as performing allocations and predictions, developing design criteria, and participating in design reviews—that both the quantitative and qualitative maintainability requirements would be satisfied [2]. The major drawback of these evaluations is that they do not reflect practical experience with the actual hardware.

Thus, it is absolutely essential to add realistic evaluations to analytical evaluations by conducting real maintainability tests and demonstrations

180

with the equipment in its operational environment. Life cycle logistic resources, such as support equipment, technical data, and technical manpower, need the same kind of evaluation. Past data on items similar to the equipment being evaluated should also be included in maintainability studies.

PLANNING AND CONTROL REQUIREMENTS FOR MAINTAINABILITY TESTING AND DEMONSTRATION

Gaining the maximum benefits from maintainability tests and demonstrations requires careful planning and control. Figure 10-1 divides requirements for planning and control into six categories [2].

Following MIL-STD-471 Guidelines

This 1966 U.S. Department of Defense document, titled *Maintainability Program Requirements,* lays out guidelines that manufacturers should carefully consider in the planning and control of maintainability demonstrations [3]. Topics covered include test conditions, selecting a test method, establishing test teams, and suggested test support materials; data collection; the pre-demonstration, formal demonstration,

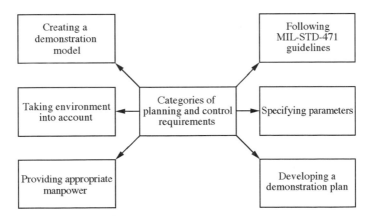

Figure 10-1. Categories of planning and control requirements for maintainability testing and demonstrations.

phases; administration, control, reporting, evaluation, and analysis procedures for the demonstration; and selection, performance, and sampling of corrective and preventive maintenance tasks. The MIL-STD-471 guidelines are considered essential for an effective maintainability demonstration.

Creating a Demonstration Model

The ideal hardware configuration for the formal test and demonstration, which will be the basis for the decision to accept or reject, would be a configuration identical to that of the final product. However, in reality this may not be always or even usually possible. The demonstration can use instead a prototype model that incorporates improvements in design, costs, and scheduling made to rectify visible shortcomings and safety hazards. The main drawback of using prototype models is that they largely consist of handcrafted parts that neatly fit together, versus the mass-produced parts that will be used in the final product. Difficulties with tolerances and other quality control problems associated with the use of mass-produced parts will therefore not show up during testing and demonstration.

A mock-up model submitted by the contractor or manufacturer can also effectively demonstrate maintainability features. Such models serve the following two basic functions:

- Providing a designer's tool for visibility, experimentation, packaging limitation, and planning, before release of the final design or drawing
- Providing a basic mechanism for demonstrating the product's proposed quantitative parameters and qualitative design features for maintainability

Providing Appropriate Manpower

The people who perform maintainability demonstrations will be essential to their success, and it is important that they possess backgrounds and skill levels similar to those of the product's final user, maintenance, and operating personnel. One way to do this is to have such personnel from the client organization perform the test. Reference 2 provides useful guidelines for selecting demonstrators.

Developing a Demonstration Plan

A good demonstration plan should cover areas such as test conditions; test planning, administration, and control; and test documentation, analysis and reporting. The plan should conform carefully to the specifics described in MIL-STD-471 [3].

During initial manufacturer or contractor participation in a program such as the validation phase, the first step is to conceive, propose, and negotiate the subject of demonstration test planning. As the program progresses, the mutually agreed-upon test plans are updated with respect to schedules, personnel selection, demonstration model designation, and identification of logistic support resource requirements. The important factor in accomplishing the demonstration on schedule and within budget is administration and control of the demonstration. Some elements of administration and control are method of organization; a team approach if desired; test monitoring; organizational interfaces; cost control; assignment of responsibilities; test event scheduling; and test data collection, reporting, and analysis.

The type and complexity of the equipment under consideration plays an important role in shaping the requirement for test documentation. Documentation requirements usually include failure reports, task selection work sheets, demonstration work sheets, frequency and distribution work sheets, demonstration task data sheets, demonstration analysis worksheets, interim demonstration reports, and final reports.

Taking Environment Into Account

Past experience has shown that equipment downtime may vary significantly between laboratory-controlled conditions and actual operational conditions. Environment is therefore an important factor in testing, and those responsible must carefully consider factors such as test facilities, support resource needs, and limitation simulations.

Specifying Parameters

The primary purpose of a formal maintainability demonstration process is verifying compliance with defined parameters. The specifications for demonstration parameters should be expressed in quantitative terms. Some examples of measurable time parameters are mean time to repair

(MTTR), mean preventive maintenance time, and mean corrective maintenance time.

TEST APPROACHES

Not all maintainability tests are formal accept/reject demonstration tests. In fact, there are many points in the product life cycle and in related maintainability program tasks that require test data, both before and after the formal decision to accept or reject. Test data may be necessary for administrative and logistic control to update corrective actions or modifications, to make decisions about maintainability design requirements, or to evaluate life cycle maintenance support. The maintainability test approaches that can provide this data fall into six categories, as shown in Figure 10-2: functional tests, dynamic tests, marginal tests, closed-loop tests, open-loop tests, and static tests.

Functional tests closely simulate normal operating conditions to establish the product's state of readiness to carry out its proposed mission effectively. The functional tests can proceed on a system-wide level

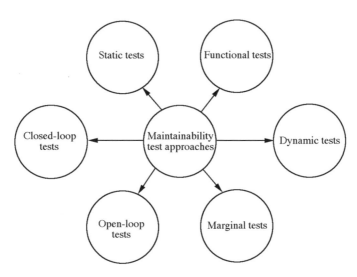

Figure 10-2. Classification of maintainability test approaches.

or focus on components such as replaceable subassemblies. These tests are performed and required at each point of evaluation for the product.

The dynamic tests simulate typical operation or uses of equipment or system so that every item involved can be checked. These tests involve a continuous input signal and analysis of the corresponding output signals to determine whether system needs are fully satisfied. Dynamic tests also provide additional information on matters such as phase characteristics, integration rates, and frequency responses.

The purpose of applying marginal tests is to isolate potential problems through the simulation of abnormal operating conditions. Such tests supply unrelated stimuli to the system or equipment under conditions such as vibrations, lowered power supply voltages, and extreme heat. Marginal testing provides greatest value as part of fault prediction, where it highlights various incipient failures resulting from abnormal operating conditions and environments.

Closed-loop tests generate information for use in evaluating design effectiveness, performance, tolerance adequacy, and other key issues. In these tests, the stimulus is adjusted on a continuous basis according to performance of the equipment or system under the test conditions. Closed-loop tests are extremely useful when a high degree of accuracy is necessary and when test points radiate performance-degrading noise levels. However, closed-loop circuits or paths in system/equipment design are very difficult to maintain and failures in the loop are difficult to diagnose.

Open-loop tests represent a refinement of dynamic and static tests. They do not provide intelligence feedback to the item being tested, that is, the stimulus is not adjusted. Open-loop tests normally provide better maintenance information than closed-loop tests because they make a direct observation of the system transfer function without the modifying influence of feedback. This method is also simpler and cheaper than closed-loop testing. It is probably the most appropriate type of testing for maintenance purposes, because it eliminates the possibility of test instability.

Static tests are simple and easy to conduct and provide information on the transient behavior of the item being tested. A series of intermittent, sequenced input signals feed into the item and, to measure its operation, the test monitors output response signals. Static tests normally establish a confidence factor but their use does not go beyond that.

TESTING METHODS

Maintainability demonstration determines whether a manufacturer or a development program has effectively satisfied qualitative and quantitative maintainability requirements. A successful maintainability demonstration depends on several factors: quality of written maintenance manuals, quality of training of repair technicians, and the quality of product design for testability. A maintainability test will not necessarily show that maintainability requirements have been met, but it does focus the manufacturer's attention on the need to meet those requirements.

MIL-STD-471 [3] is a document widely used in conducting maintainability demonstrations and presents many test methods. It describes policies and procedures for conducting maintainability demonstrations at specified points during the product development life cycle. As a single maintainability parameter can seldom address all desirable maintainability characteristics, MIL-STD-471A presents the following eleven different test methods addressing many diverse maintainability parameters [4].

The Median Equipment Repair Time Method

Maintainability demonstrations employ this method when the need or requirement is defined in terms of an equipment repair time median. The method is based on lognormally distributed corrective maintenance task times and a sample size of 20.

The Mean Method

This method is useful when the need or requirement is specified in terms of a mean value and there is a corresponding design goal value. The test plan contains two categories, Test Plan A and Test Plan B. Test Plan A assumes lognormal distribution for determining the sample size. It also assumes that a lognormal distribution can satisfactorily represent the maintenance times, and that the variance of the logarithms of the maintenance times are already known. Test Plan B makes no such assumptions. On the other hand both are fixed sample tests, and a minimum sample size of 20, that make use of the Central Limit Theorem and the asymptotic normality of the simple mean in their development.

The Man-Hour Rate (Using Simulated Faults) Method

This method demonstrates man-hour rates or man-hours per operating hour and is based on the following:

- Predicted equipment failure rate
- Total accumulative chargeable maintenance man-hours
- Total accumulative simulated demonstration operating hours

The Preventive Maintenance Times Method

This method is useful when the stated index involves mean preventive maintenance task time and/or maximum preventive maintenance task time at any percentile, and when all possible preventive maintenance tasks need to be accomplished. The test requires no allowance for assumed statistical distribution.

The Critical Maintenance Time or Man-Hours Method

This method is applicable when the need is specified in terms of:

- Required critical maintenance time or critical man-hours
- A corresponding design goal value

This test is distribution-free and can be used to establish a critical upper limit on the time or man-hours required to carry out certain maintenance tasks. The following factors are associated with this test method:

- Both the null and alternate hypotheses refer to a fixed time and the percentile varies.
- There is no need to assume the distribution of maintenance time or man-hours.

The Critical Percentile Method

Maintainability demonstrations employ this method when the requirement is defined in terms of:

- A required critical percentile
- A corresponding design goal value

If the critical percentile is fixed at 50%, then this test method is known as the test of a median. The basis for the decision criteria is the asymptotic normality of the maximum likelihood estimate of the percentile value. The method is based on two assumptions: a lognormal distribution satisfactorily describes the distribution of maintenance times, and the variance of the logarithms of the maintenance times is already known.

The Combined Mean/Percentile Requirement Method

This method is useful when the specification is stated as a dual requirement for the mean and for either the 90th or 95th percentile of maintenance times, when maintenance time is lognormally distributed.

The Percentiles and Maintenance Method

This method uses a test of proportion to demonstrate fulfillment of maximum preventive maintenance task time at any percentile, maximum corrective maintenance task time at the 95th percentile, median corrective maintenance task time, and median preventive maintenance task time, when corrective and preventive maintenance repair time distributions are unknown. The following two factors are associated with the method:

- A minimum sample size of 50 tasks is required.
- The plan holds the confidence level at 75% or 90%.

The Mean Maintenance Time and Maximum Maintenance Time Method

This method demonstrates maintainability indices such as follows:

- Mean preventive maintenance time
- Mean corrective maintenance time

- Mean maintenance time, including corrective and preventive maintenance actions

For demonstrating mean corrective maintenance time, this method's procedures are based on the Central Limit Theorem. Information on the variance of maintenance times is not needed. This allows the method to be used with any underlying distribution, provided the sample size is at least 30. The maximum maintenance time demonstration/procedure is valid for cases with lognormal underlying distribution of corrective maintenance task times.

The Chargeable Maintenance Downtime per Flight Method

This method, used in testing aircraft, makes use of the Central Limit Theorem. The chargeable downtime per flight is the allowable time, expressed in hours, for carrying out maintenance assuming that there is a specific availability and operation readiness requirement for the aircraft.

The Man-Hour Rate Method

This test method demonstrates man-hour rates, specifically man-hours per flight hour. It uses the total accumulative flight hours and the determination, during Phase II test operation, of the total accumulative chargeable maintenance man-hours. In using this test method:

- Develop appropriate ratios of equipment operating time to flight time.
- Ensure that the predicted man-hour rate pertains to flight time rather than the equipment operating time.

Reference 3 gives the statistical aspect of the various methods that have been described here.

PREPARING FOR MAINTAINABILITY DEMONSTRATIONS AND EVALUATING THE RESULTS

Figure 10-3 shows the many steps associated with the preparation for performing maintainability demonstrations and evaluating their

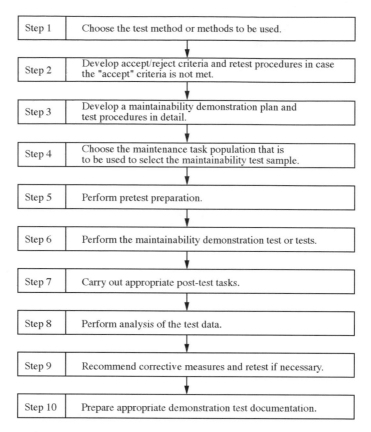

Step 1	Choose the test method or methods to be used.
Step 2	Develop accept/reject criteria and retest procedures in case the "accept" criteria is not met.
Step 3	Develop a maintainability demonstration plan and test procedures in detail.
Step 4	Choose the maintenance task population that is to be used to select the maintainability test sample.
Step 5	Perform pretest preparation.
Step 6	Perform the maintainability demonstration test or tests.
Step 7	Carry out appropriate post-test tasks.
Step 8	Perform analysis of the test data.
Step 9	Recommend corrective measures and retest if necessary.
Step 10	Prepare appropriate demonstration test documentation.

Figure 10-3. Preparation for performing maintainability demonstrations and evaluating the results.

results [5]. Step 1 is the choice of specific methods outlined in MIL-STD-471A [3]. Selection depends on factors such as product characteristics and the parameters to be demonstrated. Step 2 establishes accept/reject criteria and retest procedures in the event that the "accept" criteria is not met. Step 3 is to develop in detail a maintainability demonstration plan and test procedures. The plan addresses issues such as facility needs, manpower requirements, training needs, and documentation and equipment required. Step 4 deals with selecting maintenance task population out of which the maintainability test sample will be taken. Step 5 is pretest preparation. This includes preparing the facilities required for the test and assembling the test hardware, test support equipment, documentation, and other requirements. Step 6 is the performance of the maintainability test or tests. Step 7 involves post-test tasks, such as restoring test hardware to its original form, verifying the test hardware's acceptability for use on production items, and returning test equipment and associated facilities to pretest form. Step 8 is the analysis of test data, which includes determining whether acceptance criteria were met and analyzing the maintenance strengths and weaknesses of the product. In Step 9, corrective measures are recommended as appropriate. Step 10 deals with preparing the documentation related to the demonstration test.

To avoid pitfalls in maintainability testing [6]:

- Tailor MIL-STD-471 [3] for the program and product under consideration rather than relying on it entirely.
- Conduct some "dry run" testing, if feasible.
- Clearly define, correct, and verify all discovered deficiencies and the associated needs for corrective action.
- Conduct a new and different trial for every trial that highlights a deficiency.
- Limit the allowable trial repetitions as a requirement for cancelling the test, progressing into an "evaluate and fix" phase, and then repeating the test with newly specified faults.
- Improve the technical manual verification and validation process prior to the maintainability demonstration test.

CHECKLISTS FOR MAINTAINABILITY DEMONSTRATION PLANS, PROCEDURES AND REPORTS

Checklists play an important role in maintainability demonstration. The checklist for maintainability demonstration plans and procedures should cover [9]:

- Purpose and scope
- Test facilities
- Test requirements
- Test participation
- Test monitoring

- Test schedule
- Test conditions
- Test ground rules
- Testability demonstration considerations
- Reference documents

The purpose and scope is a statement of general test objectives and a general description of the test to be performed. Under the heading of test facilities is information such as a description of the test item's configuration, a general description of test facility, identification of the test location, test area security measures, and test safety features. Test requirements include items such as the method of generating a candidate fault list, the method of choosing and applying faults from the candidate list, the levels of maintenance to be demonstrated, a list and schedule of test reports to be issued, and support material requirements. Decisions to be made regarding test participation are the test team members, their assignments, and test decision-making authority. Test monitoring is the method of monitoring and recording test results. The test schedule should include three items: the start date, the finish date, and the test program review schedule. Two components of the test conditions are the modes of equipment operation during testing, and a description of the environmental conditions under which the test will be conducted. Under the heading of test ground rules should be a list of items to which the rules apply. These items include maintenance inspection, instrumentation failures, maintenance time limits, maintenance due to secondary failures, and technical manual usage and adequacy.

Among the components of the testability considerations are the built-in test requirements to be demonstrated, the method of selecting and simulating candidate faults, the repair levels for which requirements will be demonstrated, and acceptable levels of ambiguity at each repair level. The checklist should also detail all applicable reference documents.

The maintainability demonstration reports checklist include items related to test results, such as maintenance tasks planned, maintenance tasks selected, the selection method, measured repair times, data analysis calculation, qualifications of the personnel conducting tasks, application of the accept/reject criteria, the documentation used during maintenance, and a discussion of deficiencies identified during testing [9].

TESTABILITY

The adequacy and efficiency of a product's test and diagnostic system often influences maintainability performance [5]. The test and diagnostic system detects faults and isolates the defective unit or item. It must be reliable and its associated failures should not interfere with product performance. The following are some terms and definitions related to testability [7, 8].

- **Testability.** This is a design characteristic that makes it feasible for the operable, degraded, or inoperable status of an item to be determined and the isolation of faults within the item to be carried out effectively.
- **Built-in test.** This is product's automated capability to detect, diagnose, or isolate failures.
- **Built-in equipment.** This is the hardware that performs the built-in test function.

Three important testability characteristics of modern equipment and systems are:

- **Fault-detection capability.** This is the percentage of failures that are automatically detected and is estimated by the following relationship:

$$\theta = \frac{\lambda_d}{\lambda_s} \tag{10.1}$$

where θ is the fault detection capability.
 λ_d is the failure rate for those portions of the equipment or system where failures can be detected by the test system.
 λ_s is the system failure rate.

In maintainability demonstration the fault detection capability, θ, is:

$$\theta = \frac{\lambda_{r/s}}{\lambda_{Tr/s}} \tag{10.2}$$

where $\lambda_{r/s}$ is the number of real or simulated failures detected by the test system.
 $\lambda_{Tr/s}$ is the total number of real or simulated failures in the test.

- **Fault-isolation capability.** This parameter measures the ambiguity associated with fault-isolation activities. It can be expressed as [5]:

  - α_1 percent of the time, the system is capable of isolating a fault to within γ_1 or fewer line replacement units; and
  - α_2 percent of the time, the system is capable of isolating a fault to within γ_2 or fewer line replacement units.

The typical values for α_1, α_2, γ_1, and γ_2 are respectively, 0.90, 0.95, 1 or 2 line replacement units, and 2 or 3 line replacement units.

- **False-alarm rate.** This is the frequency with which the system indicates a failure when there is none, expressed by:

$$\lambda_f = \frac{NFA}{OT} \tag{10.3}$$

where λ_f is the false-alarm rate.

NFA is the number of false alarms experienced.

OT is the operational time interval for the system.

MIL-STD-2165A [7] presents a checklist of more than 100 testability design criteria. Some of the topics covered are:

- Test control
- Sensors
- Analog design
- Built-in testing
- Performance monitoring
- Digital design
- Mechanical design with respect to electronic functions
- Test requirements
- Test access

For example, the list provides questions about sensors such as:

- Are pressure sensors too close to pressure sensing locations to obtain wideband dynamic data?
- Are the procedures and mechanisms for sensing devices' calibration established?

Maintainability demonstration can also include demonstration of fault-detection and isolation capabilities. When real or simulated faults are used to demonstrate product maintainability, the recorded fault detection and isolation activities and times also demonstrate the level of adherence to testability requirements. During normal test or operational activities, it is not possible to simulate false alarms. Thus the measurement of this parameter depends on any false alarms that happen to occur during a maintainability demonstration program and on their subsequent frequency of occurrence.

MAINTAINABILITY DATA

Failure and repair data are invaluable in reliability and maintainability studies. The uses of such data include determining product maintenance needs, predicting product maintainability, performing life-cycle cost studies, and determining product replacement policies. There are many sources of such data. For example, during the product life cycle, data sources include reports generated by the repair facility, past experience with similar or identical items, failure reporting systems developed and used by customers, and warranty claims. Some other sources for obtaining reliability and maintainability data are [9, 10, 11]:

- Government Industry Data Exchange Program (GIDEP), GIDEP Operation Center Corona, California.
- Reliability Analysis Center, Rome Air Development Center, Griffiss Air Force Base, Rome, New York, 13440-8200.
- Maintenance and Operational Data, Access System (MODAS), Air Force Logistics Command/MMTS, Wright-Patterson Air Force Base, Ohio.
- Coogan, F. C. "RAMS Data Bank for Electrical Power Equipment." *Proceedings of the Inter-Ram Conference for Electrical Power Industry,* June 1986, pp. 306–310.
- English, C. "In-Service Reliability Estimates from Maintenance Data." *Proceedings of the National Reliability Conference,* April 1987, pp. 5 C/4/1-5C/4/7.
- Erto, P. "Reliability Assessments by Repair Shops via Maintenance Data." *Proceedings of the National Reliability Conference,* April 1987, pp. 5 C/3/1-5C/3/12.

- *Reliability and Maintainability Data for Industrial Plants,* Report No. TD-84-3, A. P. Harris and Associates, Ottawa, Canada, 1984.
- Sherwin, D. J. "Improved Schedules by Using Data Collected Under Preventive Maintenance." *IEEE Transactions of Reliability,* Vol. 33, 1984, pp. 315–320.
- Sherwin, D. J. and Lees, F. P. "An Investigation of the Application of Failure Data Analysis to Decision-Making in Maintainability of Process Plant." *Transactions of the Institute of Mechanical Engineers,* 1980, pp. 301–319.
- Thunstedt, B. "Use of Maintenance Management Systems for Collection of Reliability and Maintainability Data." *Proceedings of the VTT Symposium on Reliability Data Collection and Validation,* Technical Research Center of Finland, VTT Symposium 32, October 1982, pp. 42–53.
- Weal, T. W. "Aviation Maintenance Data Collection in the U.S. Navy." *Proceedings of the Reliability and Maintainability Conference,* July 1965, pp. 839–861.

PROBLEMS

1. List the six categories of planning and control requirements for maintainability testing and demonstrations. Describe two categories in detail.
2. Describe the following two maintainability test approaches:
   - Functional tests
   - Marginal tests
3. Discuss the following three test methods described in MIL-STD-471:
   - The medium equipment repair time method
   - The mean method
   - The preventive maintenance times method
4. Describe the steps in preparing for and performing a maintainability demonstration and in evaluating its results.
5. Discuss ways to avoid pitfalls in maintainability testing.
6. Define the following three terms:
   - Testability
   - Built-in test equipment
   - Built-in test

7. Describe testability demonstration in detail.
8. Discuss the uses of maintainability data.
9. What are the benefits of performing a maintainability demonstration?
10. Define the following two parameters associated with testability:
    * Fault-detection capability
    * Fault-isolation capability

REFERENCES

1. MIL-STD-470, *Maintainability Program Requirements*. Department of Defense, Washington, D.C., 1966.
2. AMCP-706-133, *Maintainability Engineering Theory and Practice*. Department of Defense, Washington, 1976.
3. MIL-STD-471, *Maintainability Verification/Demonstration/Evaluation*. Department of Defense, Washington, D.C., 1966. (Revision "A," 1973.)
4. PRIM-1, *A Primer for DOD Reliability, Maintainability and Safety Standards*. Reliability Analysis Center, Rome Air Development Center, Griffiss Air Force Base, New York, 1988.
5. Grant Ireson, W., Coombs, C. F. and Moss, R. Y. *Handbook of Reliability Engineering and Management*. McGraw-Hill Inc., New York, 1996.
6. Bentz, R. W. "Pitfalls to Avoid in Maintainability Testing." *Proceedings of the Annual Reliability and Maintainability Symposium*, 1982, pp. 278–282.
7. MIL-STD-2165A, *Testability Program for Systems and Equipment*. Department of Defense, Washington, D.C., 1993.
8. Dhillon, B. S. and Viswanath, H. C. "Bibliography of Literature on Testability." *Microelectronics and Reliability*, Vol. 30, 1990, pp. 375–415.
9. *RADC Reliability Engineer's Toolkit*. Systems Reliability and Engineering Division, Rome Air Development Center, Griffiss Air Force Base, New York, 1988.
10. Dhillon, B. S. and Viswanath, H. C. "Bibliography of Literature on Failure Data." *Microelectronics and Reliability*, Vol. 30, 1990, pp. 723–750.
11. Dhillon, B. S. *Mechanical Reliability: Theory, Models, and Applications*. American Institute of Aeronautics and Astronautics, Washington, D.C., 1988.

Maintenance Models and Warranties

INTRODUCTION

As mentioned earlier, maintenance and maintainability are not the same but are closely interlinked [1]. Over the years, many mathematical models have been developed to better define and predict aspects of maintenance. There has been a similar effort to develop models for warranties. Usually, customers of engineering products place emphasis on both maintenance and warranties in making their procurement decisions. A warranty on manufactured goods spells out the manufacturer's responsibility in case the goods are defective. One study of 369 United States manufacturers found that more than 95% had written warranties of some sort on their products. The average cost of a warranty claim was approximately 2% of sales [2].

MAINTENANCE MODELS

Various kinds of mathematical models are available to assist in making decisions concerning product maintenance.

Maintenance Model I

This model determines the optimum number of inspections per facility per unit of time. An inspection is often disruptive, but it usually decreases downtime because it means fewer breakdowns. References 3 and 4 present mathematical models that calculate optimum number of inspections with minimum total downtime of the equipment. The total downtime is expressed by

$$T_D = xT_{pf} + \frac{kT_{Bf}}{x} \tag{11.1}$$

where T_D is the total downtime per unit of time for a facility.
   x is the number of inspections per facility per unit of time.
   k is a constant for a specific facility.
   T_{pf} is the downtime per inspection for a facility.
   T_{bf} is the downtime per breakdown for a facility.

By taking derivatives of Equation 11.1 with respect to x we get

$$\frac{dT_D}{dx} = T_{pf} - kT_{Bf}x^{-2} \tag{11.2}$$

Setting Equation 11.2 equal to zero and then rearranging it results in

$$x^* = \left(\frac{kT_{Bf}}{T_{pf}} \right)^{1/2} \tag{11.3}$$

where x^* is the optimum number of inspections per facility per unit of time.

Substituting Equation 11.3 into Equation 11.1 leads to

$$T_D^* = 2(kT_{pf}T_{Bf})^{1/2} \tag{11.4}$$

where T_D^* is the optimum total downtime per unit of time for a facility.

Example 11-1

   Assume that the following data are associated with a piece of engineering equipment:

$T_{pf} = 0.009$ month

$T_{Bf} = 0.2$ month

$k = 2$

Determine the optimum number of inspections per month using Equation 11.3.

Substituting this data into Equation 11.3 yields

$$x^* = \left(\frac{2(0.2)}{0.009} \right)^{1/2} = 6.67$$

The optimum number of inspections per month for the piece of engineering equipment is 6.67.

Maintenance Model II

This model determines the optimum time interval between replacements. The goal is to minimize average annual total cost with respect to the time between replacements or the life of the equipment in years. The average cost consists of three elements: mean investment cost, mean maintenance cost, and mean operating cost. The total average cost is

$$C_T = OC_1 + MC_1 + \frac{IC}{x} + \left(\frac{x-1}{2} \right)(\alpha_{oc} + \alpha_{mc}) \qquad (11.5)$$

where C_T is the average total cost.

   x is the equipment life expressed in years.

   OC_1 is the equipment's operational cost for the first year.

   MC_1 is the equipment's maintenance cost for the first year.

   IC is the cost of investment.

   α_{oc} is the amount by which operational cost increases per year.

   α_{mc} is the amount by which maintenance cost increases per year.

Differentiating Equation 11.5 with respect to x leads to

$$\frac{dC_T}{dx} = \frac{1}{2}(\alpha_{oc} + \alpha_{mc}) - \frac{IC}{x^2} \tag{11.6}$$

Setting Equation 11.6 equal to zero and rearranging it results in

$$x^* = \left(\frac{2IC}{\alpha_{oc} + \alpha_{mc}}\right)^{1/2} \tag{11.7}$$

where x^* is the optimum replacement interval.

Substituting Equation 11.7 into Equation 11.5 leads to

$$C_T^* = OC_1 + MC_1 - \left(\frac{\alpha_{oc} + \alpha_{mc}}{2}\right) + [2IC(\alpha_{oc} + \alpha_{mc})]^{1/2} \tag{11.8}$$

where C_T^* is the minimum average annual total cost.

Example 11-2

The following data apply to an engineering system:

$\alpha_{oc} = \$1,000$, $\alpha_{mc} = \$500$, $IC = \$50,000$

Determine the optimum replacement interval using Equation 11.7. Substituting the given data into Equation 11.7 yields

$$x^* = \left(\frac{2(50,000)}{1,000 + 500}\right)^{1/2} = 8.16 \text{ years}$$

The optimum replacement period for the engineering system is 8.16 years.

Maintenance Model III

This model determines the optimum test interval for engineering systems, especially nuclear safety systems. Such systems are tested periodically to determine their operational readiness. A shorter time interval between tests will in one regard lead to higher availability of the system because failures will be detected earlier [5]. However, testing may require shutdown of an operating system. Under such conditions, it is necessary to establish some sort of optimum period between tests. For the purpose of developing a mathematical model to determine the optimum period between tests, Figure 11-1 shows one test cycle of the safety system.

The assumptions associated with this model are [6]:

- The safety system availability is equal to unity immediately after each test.
- The safety system availability is equal to zero during each test.
- The safety system availability decreases exponentially until the start of the next test.
- The safety system failure rate is constant.

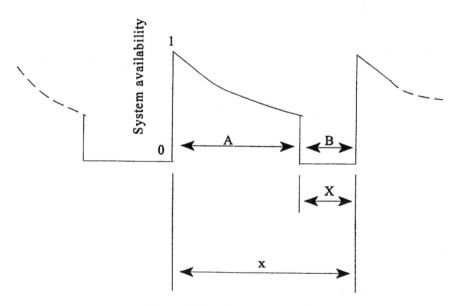

Figure 11-1. Test interval cycle.

The following symbols are associated with this maintenance model:

λ is the constant failure rate of the safety system.

A   is the time domain as indicated in Figure 11-1.

B   is the time domain as indicated in Figure 11-1.

X   is the time required to perform a test.

x   is the time between tests.

P(C)   is the probability that the safety system is in operational readiness state.

P(A)   is the probability that the future random point lies in A.

P(B)   is the probability that the future random point lies in B.

P(C/B) is the probability that the safety system is in operational readiness state given that the future random point lies in B.

P(C/A) is the probability that the safety system is in operational readiness state given that the future random point lies in A.

The probability of the safety system being in operational readiness state is expressed by

$$P(C) = P(A)P(C/A) + P(B)P(C/B) \tag{11.9}$$

Since the safety system is unavailable during the test interval,

$$P(C/B) = 0 \tag{11.10}$$

Using Figure 11-1, we write

$$P(A) = \frac{x - X}{x} \tag{11.11}$$

P(C/A) can be obtained by averaging reliability over the time interval as follows:

$$P(C/A) = \frac{1}{t} \int_{0}^{t} e^{-\lambda y} dy \tag{11.12}$$

where $t = x - X$.

Thus,

$$P(C/A) = \frac{1 - e^{-\lambda t}}{\lambda t} \tag{11.13}$$

Since $P(C/B) = 0$, Equation 11.9 reduces to

$$P(C) = P(A)P(C/A) \tag{11.14}$$

Substituting Equations 11.11 and 11.13 into Equation 11.14 yields

$$P(C) = \left(\frac{x - X}{x}\right)\frac{(1 - e^{-\lambda t})}{\lambda t} = \frac{t}{x}\frac{(1 - e^{-\lambda t})}{\lambda t} = \left(\frac{(1 - e^{-\lambda t})}{\lambda t}\right) \tag{11.15}$$

Differentiating Equation 11.15 with respect to x, and then equating it to zero and using the $t = x - X$ relationship, leads to

$$e^{-\lambda t}(1 + \lambda t) - e^{-\lambda X} = 0 \tag{11.16}$$

To obtain an approximate formula for x, we use the following approximations for the exponential function:

$$e^{-\lambda t} \cong 1 - \lambda t + \frac{\lambda^2 t^2}{2} \tag{11.17}$$

and

$$e^{-\lambda X} \cong 1 - \lambda X + \frac{\lambda^2 X^2}{2} \tag{11.18}$$

Thus, substituting Equations 11.17 and 11.18 into Equation 11.16 yields

$$x^2 = \frac{2X}{\lambda} - X^2 \tag{11.19}$$

Note that to obtain Equation 11.19, the term $\lambda^2 t^3/2$ was set equal to zero. If X is much smaller than λ^{-1} in Equation 11.19, then the optimum value x* is given by

$$x^* = \left(\frac{2X}{\lambda} \right)^{1/2} \qquad (11.20)$$

Thus, the optimum time between tests can be calculated using Equation 11.20.

Maintenance Model IV

This model is concerned with a parallel system composed of k identical machines or pieces of equipment, with output fed into the next stage of the production process. For the system success at least one machine must function normally. Furthermore, the total cost involved in system operation and downtime losses with respect to k is minimized [7].

Using queuing theory knowledge, we write the following relationship to obtain the average proportion of unit of time that the parallel system is unavailable:

$$UA = \left(\frac{\lambda}{\lambda + \mu} \right)^k \qquad (11.20)$$

where UA is the system unavailability.

λ is the constant machine failure rate.

μ is the constant machine repair rate.

Thus, the total cost, TC, is

$$TC = (UA)(DC) + k(MOC) \qquad (11.21)$$

where DC is the downtime cost per unit of time.

MOC is the single machine's operational cost per unit of time.

Substituting Equation 11.20 into Equation 11.21 yields

$$TC = \left(\frac{\lambda}{\lambda + \mu} \right)^k (DC) + k(MOC) \qquad (11.22)$$

Differentiating Equation 11.22 with respect to k leads to

$$\frac{d(TC)}{dk} = \left[\left(\frac{\lambda}{\lambda + \mu} \right)^k \ln \left(\frac{\lambda}{\lambda + \mu} \right) \right] (DC) + MOC \qquad (11.23)$$

By setting Equation 11.23 equal to zero and then solving it for k, we get

$$k^* = \left(\frac{\ln \left[- \dfrac{MOC}{(DC) \ln UA_1} \right]}{\ln UA_1} \right) \qquad (11.24)$$

where k* is the optimal number of machines to be used in the parallel configuration for minimum total cost.

$$UA_1 = \frac{\lambda}{\lambda + \mu} \qquad (11.25)$$

Example 11-3

The following data apply to an engineering system used to produce certain mechanical parts:

$\lambda = 4$ failures per month, $\mu = 10$ repairs per month, MOC = $150, and DC = $1,500

Using Equation 11.24, determine the optimum number of machines to be used in the parallel configuration to minimize total cost.
Inserting the specified data into Equation 11.25 leads to

$$UA_1 = \frac{4}{4 + 10} = 0.2857$$

Using this result and the other data in Equation 11.24, we get

$$k^* = \left(\frac{\ln\left[-\dfrac{150}{(1,500)\ln(0.2857)} \right]}{\ln(0.2857)} \right) = 2.018 \cong 2 \text{ machines}$$

Thus, using two machines in the parallel configuration minimizes the total cost.

Maintenance Model V

This model determines the optimum replacement time for an item under ordinary periodic replacement policy. Under this policy, an item is replaced with a new one every x_p accumulated hours of operation [7, 8]. If the item malfunctions prior to x_p hours, it is repaired only minimally so that its instantaneous failure rate, $\lambda(x)$, corresponding to its probability density function, $f(x)$, remains the same as it was before failure. It is assumed that each failure is detected instantaneously and the minimum repair time is negligible.

The cost function for model is expressed by

$$K = \frac{C_{pr} + C_{mr}E[\alpha(x_p)]}{x_p} \tag{11.26}$$

where k is the cost per unit per operating hour.

C_{mr} is the cost of minimal repair.

C_{pr} is the cost associated with planned preventive replacement.
$E[\alpha(x_p)]$ is the expected number of failures followed by minimal repair activity during an interval x_p.

Thus we have

$$E[\alpha(x_p)] = \int_0^{x_p} \lambda(x)dx \tag{11.27}$$

where

$$\lambda(x) = \frac{f(x)}{R(x)} \tag{11.28}$$

where f(x) is the failure probability density function of a unit.
R(x) is the reliability function of a unit.
λ(x) is the time dependent failure rate of unit.

Substituting Equation 11.27 into Equation 11.26 results in

$$K = \frac{C_{pr} + C_{mr} \int_0^{x_p} \lambda(x)dx}{x_p} \tag{11.29}$$

Example 11-4

A mechanical unit receives preventive maintenance per the policy described, and the Rayleigh probability density function expresses its times to failures as follows:

$$f(x) = \frac{2x}{(20)(20)} e^{-\left(\frac{x}{20}\right)^2} \tag{11.30}$$

Assume that the planned preventive replacement cost is $10 and the minimal repair cost is $50. Calculate the optimum preventive replacement time.

By integrating Equation 11.30 over the time interval [o, x], we get

$$F(x) = \int_0^x \frac{2x}{(20)^2} e^{-\left(\frac{x}{20}\right)^2} dx = 1 - e^{-\left(\frac{x}{20}\right)^2} \tag{11.31}$$

where F(x) is the cumulative distribution function.

Subtracting Equation 11.31 from unity leads to

$$R(x) = 1 - \left[1 - e^{-\left(\frac{x}{20}\right)^2} \right] = e^{-\left(\frac{x}{20}\right)^2} \tag{11.32}$$

where $R(x)$ is the reliability at time x.

By inserting Equations 11.30 and 11.31 into Equation 11.28, we get

$$\lambda(x) = \frac{2x}{(20)^2} \tag{11.33}$$

Substituting Equation 11.33 into Equation 11.27 yields

$$E[\alpha(x_p)] = \int_0^{x_p} \frac{2x}{(20)^2} \, dx = \left(\frac{x_p}{20}\right)^2 \tag{11.34}$$

By inserting Equation 11.34 into Equation 11.26, we get

$$K = \frac{C_{pr} + C_{mr}\left(\dfrac{x_p}{20}\right)^2}{x_p} \tag{11.35}$$

Differentiating Equation 11.35 with respect to x_p, and then setting the resulting equation equal to zero, leads to

$$\frac{dK}{dx_p} = -\frac{C_{pr}}{x_p^2} + \frac{C_{mr}x_p}{(20)^2} = 0 \tag{11.36}$$

Solving Equation 11.36 leads to the following optimal replacement time:

$$x_p^* = (20)\left(\frac{C_{pr}}{C_{mr}}\right)^{1/2} \tag{11.37}$$

where x_p^* is the optimum replacement time.

By inserting the specified data into Equation 11.37 we get

$$x_p^* = (20)\left(\frac{10}{50}\right)^{1/2} = 8.94 \text{ hours}$$

The optimum preventive replacement time for the mechanical unit is 8.94 hours.

Maintenance Model VI

This model is similar to Model V except that in this case the objective is to minimize the total downtime per unit of time—in other words, to minimize equipment unavailability. The model represents the constant interval replacement policy. Two important factors associated with this policy are [9]:

- Replacements are carried out at predetermined times irrespective of the age of the equipment or unit being replaced.
- Replacements are performed when equipment fails.

Total equipment downtime per unit of time, $DT(x_p)$, is [10, 11]

$$DT(x_p) = \frac{TDT}{CL} \qquad (11.38)$$

where TDT is the total downtime of the equipment under consideration.
     CL is the cycle length or the length of the preventive replacement cycle.

In turn, TDT is expressed by

$$TDT = DTF + DTPR \qquad (11.39)$$

where DTF is the equipment downtime due to failure.
     DTPR is the equipment downtime due to preventive replacement.

Alternatively, TDT can also be expressed by

$$TDT = (ENF)(TPR) + X_{pr} \qquad (11.40)$$

where ENF is the expected number of failures in time interval $[0, x_p]$.
TPR is the time to perform a failure replacement.
X_{pr} is the time to perform a preventive replacement.

The cycle length, CL, is given by

$$CL = X_{pr} + x_p \qquad (11.41)$$

Inserting Equations 11.40 and 11.41 into Equation 11.38 yields

$$DT(x_p) = \frac{(ENF)(TPR) + X_{pr}}{X_{pr} + x_p} \qquad (11.42)$$

The optimal value of x_p may be obtained in similar fashion to that for Maintenance Model V.

WARRANTIES

Increasing competition forces manufacturers to continually add new features to their products to increase or retain their market shares. Some examples of these features are lower prices, better quality, and attractive warranties. Many manufacturers cite the following basic reasons for providing warranties on their products [12]:

- To compete more effectively
- To define liabilities
- To support their manufactured products

More specific reasons for offering a warranty are [13]:

- It enhances the changes of prompt acceptance of the manufactured product.
- It helps to increase the manufactured product's marketability.

- It assures the customers that their newly procured products will at least meet contractual requirements.
- It helps expedite payment for the product from the purchaser.
- It encourages manufacturers to produce high-quality, reliable products.
- It forces the manufacturers to provide better product support documentation and services.
- It places the responsibility for rectifying any faults with the product manufacturer.

In general, warranties fall into the three categories [9] shown in Figure 11-2.

Under an unlimited free replacement warranty, whenever a product fails prior to the end of the warranty length, it is replaced or repaired to its original condition at no charge to the customer. In turn, each replacement item is covered by a warranty identical to the original

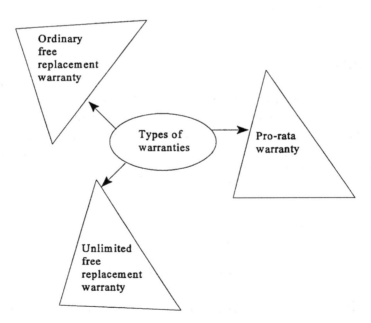

Figure 11-2. Common kinds of warranties.

purchase warranty. This type of warranty is usually used for small electronic appliances because of their high early failure rates and is normally restricted to a short span of time.

The ordinary free replacement warranty is basically the same as the unlimited free replacement warranty, with one exception. The replacement or repaired item is covered by an ordinary free replacement warranty only during the remaining period of the original warranty. This is probably the most widely used type of warranty and often covers durables such as kitchen appliances and cars [14]. It must be remembered that with either of these two types of warranties, a lengthy warranty period will lead to a very high warranty cost.

Under a pro-rata warranty, if the product malfunctions prior to the end of the warranty period, it is replaced at a cost that is a function of the item's age at the time of failure. The replacement product is covered by an identical warranty. From the manufacturer's point of view, this type of warranty is more beneficial.

In planning a warranty program, manufacturers must decide on the type of warranty policy, the duration of the warranty period, and the amount of capital to be set aside to cover potential expenses for failures during the warranty period.

This section presents some of the mathematical warranty models that have been developed.

Warranty Model I

This model estimates a manufacturer's warranty cost [15]. The warranty cost of an item is

$$C_w = FC_w + \lambda(T_O) \, AC_w \tag{11.43}$$

where C_w is the warranty cost of an item.

AC_w is the average cost to the manufacturer of the item.

FC_w is the warranty's fixed cost to the manufacturer.

T_O is the total operating time, in hours, of the item over its warranty period.

λ is the constant failure rate of the item, expressed in failures per hour.

Warranty Model II

This model determines the warranty reserve fund when the product times to failure are exponentially distributed [9, 16]. The symbols associated with this model are:

λ is the constant failure rate.
α is the product lot size for warranty reserve determination.
T_w is the warranty period duration.
t is time.
E is the expected number of failures at time t.
θ is the constant unit product price, including warranty cost.
F is the total warranty reserve fund for α number of units.
k is the warranty reserve cost per unit of product and is given by
   $k = F/\alpha$.

The pro-rata customer rebate at time t is expressed by

$$CR(t) = \theta[1 - (t/T_w)], \text{ for } 0 < t < T_w \tag{11.44}$$

For exponentially distributed times to failure of a product, the probability of failure is [6]:

$$F(t) = 1 - e^{-\lambda t} \tag{11.45}$$

The expected number of failures occurring at any time t is expressed by

$$E[n(t)] = \alpha F(t) = \alpha(1 - e^{-\lambda t}) \tag{11.46}$$

The total number of failures in the interval [t, t + dt] is

$$dE[n(t)] = \frac{\partial E[n(t)]}{\partial t}\, dt = [\alpha \lambda e^{-\lambda t}]dt \tag{11.47}$$

The cost for the failures in interval [t, t + dt] is given by

$$d(F) = CR(t)dE[n(t)] = \theta\left[1 - \left(\frac{t}{T_w}\right)\right]\alpha \lambda e^{-\lambda t}dt \tag{11.48}$$

The total cost for all failures occurring in time interval $[0, T_w]$ is

$$F = \int_0^{T_w} \alpha\lambda\theta\left[1 - \left(\frac{t}{T_w}\right)\right]e^{-\lambda t}dt = \alpha\theta\left[1 - \left(\frac{1}{\lambda T_w}\right)(1 - e^{-\lambda T_w})\right] \qquad (11.49)$$

The warranty reserve cost per unit of product is

$$k = \frac{F}{\alpha} = \theta\left[1 - \left(\frac{1}{\lambda T_w}\right)(1 - e^{-\lambda T_w})\right] \qquad (11.50)$$

By rearranging Equation 11.50, we get

$$\frac{k}{\theta} = 1 - \left(\frac{1}{\lambda T_w}\right)(1 - e^{-\lambda T_w}) \qquad (11.51)$$

Suppose the constant unit product price consists of two components as follows:

$$\theta = \theta' + k \qquad (11.52)$$

where θ' is the unit product price excluding the warranty cost.

By rearranging Equation 11.52, we get

$$\theta' = \theta\left(1 - \frac{k}{\theta}\right) \qquad (11.53)$$

or

$$\theta = \frac{\theta'}{\left(1 - \frac{k}{\theta}\right)} \qquad (11.54)$$

Example 11-5

A manufacturer of washing machines plans to provide a 12-month warranty on its new model. The estimated failure rate of the new model is 0.005 failures per month. The approximate manufacturer's cost, excluding the warranty cost, of the new washing machine is $400. Calculate the warranty reserve fund for a production run of 500 washing machines.

Inserting the specified data into Equation 11.52 yields

$$\frac{k}{\theta} = 1 - \left[\frac{1}{(0.005)(12)} \right] [1 - e^{-(0.005)(12)}] = 0.0294$$

By substituting this result and the other specified data into Equation 11.54, we get

$$\theta = \frac{400}{(1 - 0.0294)} = \$412.12$$

Thus, using Equation 11.49, we get the following amount for the warranty reserve fund for the production run of 500 washing machines:

$$F = (500)(412.12)(0.0294) = \$6,058.16$$

Warranty Model III

This model is an extension of Warranty Model II. It is used when administrative costs and the errors in estimating pro-rata claims are too high and the manufacturer thus opts for an alternative warranty plan, such as paying a fixed or lump-sum rebate to the buyer for any failure occurring during the duration of the warranty. Again, in this case we are concerned with determining the following two factors:

• The product's adjusted price
• The warranty reserve fund needed to meet claims made by customers

Thus, we assume that proportion p of the unit cost will be refunded as a lump sum and that the unit lump-sum rebate is expressed by [16, 9]

$$\beta = p\theta \qquad (11.55)$$

Substituting $CR(t) = p\theta$ into Equation 11.48 gives a total cost for all failures occurring in time interval $[0, T_w]$

$$F = \int_0^{T_w} p\theta\alpha\lambda e^{-\lambda t} dt = p\theta\alpha(1 - e^{-\lambda T_w}) \tag{11.56}$$

Rearranging Equation 11.56 yields

$$k_s = \frac{F}{\alpha} = p\theta(1 - e^{-\lambda T_w}) \tag{11.57}$$

where k_s is the warranty reserve cost per unit of product under the new lump-sum rebate plan.

To find an equal warranty reserve fund per unit under both the pro-rata and the lump-sum plans, we equate Equations 11.50 and 11.57

$$\theta\left[1 - \left(\frac{1}{\lambda T_w}\right)(1 - e^{-\lambda T_w})\right] = p\theta(1 - e^{-\lambda T_w})$$

$$\left[1 - \left(\frac{1}{\lambda T_w}\right)(1 - e^{-\lambda T_w})\right] = p(1 - e^{-\lambda T_w}) \tag{11.58}$$

Solving Equation 11.58 for p results in

$$p = \frac{1}{(1 - e^{-\lambda T_w})} - \frac{1}{\lambda T_w} \tag{11.59}$$

Example 11-6

The manufacturer in Example 11-5 adopts a warranty plan that provides a lump sum of the initial procurement price to customers whose washing machines malfunction before the warranty expires. Calculate the portion of the price to be refunded. Also prove that the total warranty reserve fund for both the lump-sum and pro-rata plans is the same.

Inserting the data given into Equation 11.59 yields

$$p = \frac{1}{[1 - e^{-(0.005)(12)}]} - \frac{1}{(0.005)(12)} = 0.5049$$

This means the washing machine manufacturer should refund approximately half the initial price to its customers when the washing machines malfunction during the warranty period.

The cost of the washing machine, including the warranty cost, is

$$\theta = \theta' + k = \$412.12$$

Substituting these two results into Equation 11.55 yields

$$\beta = (0.5049)(412.12) = 208.08$$

Inserting this result and the other specified data into Equation 11.56 yields

$$F = (208.08)(500)[1 - e^{-(0.005)(12)}] = \$6,058.16$$

This is identical to the result obtained in Example 11-5. It proves that the total warranty reserve fund for both plans is same.

Warranty Model IV

This model determines warranty costs for repairable products. When a product fails, the company performs minimum repair necessary to restore it to working condition. After the repair, there is no warranty extension provided [9, 17]. The symbols associated with this model are:

Ac_r	is the average cost per repair.
i	is the nominal interest rate for discounting the future cost associated with the product.
T_w	is the warranted period.
t	is time.

λ   is the scale parameter of Weibull distribution.

b   is the shape parameter of Weibull distribution.

λ(t)   is the hazard rate or time dependent failure rate associated with the product.

H(t)   is the cumulative hazard function and is given by $\int_0^t \lambda(t)dt$.

C_{pw} is the present worth of repair cost during the warranty period.

C   is the present worth of repair for an item/product having lifetime warranty.

poim (i, μ)   is the Poisson probability mass function (pmf), given by $e^{-\mu}\mu^i/i!$.

For minimal repair upon failure, the hazard rate resumes at λ(t) instead of returning to λ(0). The product's failure times are thus not renewal points but can be represented by a nonhomogeneous Poisson process. The probability density function of the time to the kth failure is therefore

$$f_k(t) = \lambda(t) \text{ poim}[k - 1; H(t)]$$

or

$$f_k(t) = \lambda b(\lambda t)^{b-1}\left[\frac{e^{-(\lambda t)^b}(\lambda t)^{(k-1)b}}{\Gamma(k)}\right] \qquad (11.60)$$

for a Weibull probability density function where $H(t) = (\lambda t)^b$.

The present value of the cost of the repairs occurring during the warranty time interval $[0, T_w]$ is

$$C_{pw} = \sum_{k=1}^{\infty} \int_0^{T_w} AC_r e^{-it}f_k(t)dt = AC_r b \int_0^{\lambda T_w} e^{-ix/\lambda}x^{b-1}dx$$

$$= AC_r b(\lambda T_w)^b e^{iT_w} \sum_{j=0}^{\infty} \frac{(iT_w)^j}{b(b+1)\ldots(b+j)} \qquad (11.61)$$

Equation 11.61 can also be rewritten as

$$C_{pw} = AC_r b(\lambda/i)^b e^{iT_w} \sum_{j=0}^{\infty} \frac{(iT_w)^{b+j}}{\beta(\beta+1) \ldots (\beta+j)} \tag{11.62}$$

The present value of repair for a product having lifetime warranty is

$$C_{\infty} = AC_r \int_0^{\infty} e^{-it} \lambda(t) dt = AC_r \left(\frac{\lambda}{i}\right)^b \Gamma(b+1) \tag{11.63}$$

where $\Gamma(b + 1) = b!$

Example 11-7

Assume that the constant failure rate, λ, of an electronic equipment part is 2 failures per year and the average cost per repair is $500. If the nominal annual interest rate is 8%, determine the expected warranty cost both for 12 months and lifetime for the part.

Per the data given, inserting $b = T_w = 1$ into Equation 11.62 yields

$$C_{pw} = AC_r(\lambda/i)e^{-i}(e^i - 1) \tag{11.64}$$

Substituting the other given data into Equation 11.64, we get

$$C_{pw} = (500)[2/0.08][1 - e^{-0.08}] = \$961.04$$

Inserting the data into Equation 11.63 leads to

$$C_{\infty} = (500)\left(\frac{2}{0.08}\right)(2) = \$25,000$$

The expected warranty cost for 12 months is $961.04, and the lifetime warranty cost is $25,000.

Warranty Model V

This model estimates the life cycle cost of avionics equipment under warranty. With some modifications, the same model also applies to other equipment. The life cycle cost is [17]:

$$\text{LCC}_{ww} = (1 + MF)(RF)[RMC + WRC]$$
$$+ DMC + T(RSC) + n(PC)(AF)$$
$$\tag{11.65}$$

$$RF \equiv (1 + i)^{T/12}, \quad 0 < i < 1 \tag{11.66}$$

where LCC_{ww} is the life cycle cost of avionics equipment under warranty.

MF is the overhead fee charged by the manufacturer.

RF is the risk factor the manufacturer uses to calculate cost for a warranty interval of T months.

RMC is the expected amortized cost of reliability-related modifications.

WRC is the direct warranty repair cost to manufacturer.

DMC is the user's direct maintenance cost.

T is the time period in months.

RSC is the recurring support cost per month.

n is the total number of units procured.

PC is the purchasing price of each unit.

AF is the amortization factor for time interval [0, T] and is expressed by T divided by the expected life of equipment.

PROBLEMS

1. Write an essay on maintenance models.
2. Define the term "warranty." What are the specific reasons for providing warranties?
3. Describe the following classifications of warranties:
   - Ordinary free replacement warranty
   - Pro-rata warranty
   - Unlimited free replacement warranty
4. Assume that in Maintenance Model II, the values of the associated parameters are as follows:

$$\alpha_{oc} = \$1,200, \ \alpha_{mc} = \$500, \ IC = \$65,000$$

Determine the optimum replacement interval. Comment on the end result.

5. Assume that an engineering system receives preventive maintenance per the policy described in Maintenance Model V. The times to failure of the system are described by the Weibull probability density function as follows:

$$f(y) = \frac{2}{(30)^2} \, ye^{-\left(\frac{y^2}{900}\right)}$$

The planned preventive replacement cost and the minimum repair cost are $20 and $70, respectively. Determine the optimum preventive replacement time.

6. A manufacturer of television sets intends to provide a one-year warranty (per Warranty Model II) on its new model. The times to failure of the new model are exponentially distributed, with a failure rate of 0.001 failures per month. The estimated manufacturer's cost, excluding the warranty cost, of the new television set is $700. Determine the warranty reserve fund for a production run of 600 television sets.

7. Assume that in Question 6, the manufacturer adopts an equivalent lump-sum warranty plan instead of the pro-rata plan. Compute the portion of the purchase price cost to be refunded to customers. Also, prove that the total warranty reserve fund for both the plans is same.

REFERENCES

1. Patton, J. D. *Maintainability and Maintenance Management.* Instrument Society of America, Research Triangle Park, North Carolina, 1980.
2. McGuire, E. P. *Industrial Product Warranties: Policies and Practices.* The Conference Board, New York, 1980.
3. Wild, R. *Essential of Production and Operations Management.* Holt, Rinehart and Winston, London, 1985.
4. Dhillon, B. S. *Mechanical Reliability: Theory, Models and Applications.* American Institute of Aeronautics and Astronautics, Inc., Washington D.C., 1988.
5. Jacobs, I. M. "Reliability of Engineered Safety Features as a Function of Testing Frequency." *Nuclear Safety,* Vol. 9, 1968, pp. 303–312.
6. Dhillon, B. S. *Power System Reliability, Safety, and Management.* Ann Arbor Science Publishers, Ann Arbor, Michigan, 1983.

7. Barlow, R. W. and Proschan, F. *Mathematical Theory of Reliability.* John Wiley & Sons, New York, 1965.
8. Kececioglu, D. *Maintainability, Availability and Operational Readiness Engineering.* Prentice-Hall, Inc., New Jersey, 1995.
9. Elsayed, E. A. *Reliability Engineering.* Addison Wesley Longman, Inc., Reading, Massachusetts, 1996.
10. Jardine, A. K. S. *Maintenance, Replacement, and Reliability.* John Wiley & Sons, New York, 1973.
11. Blanks, H. S. and Tordan, M. J. "Optimum Replacement of Deteriorating and Inadequate Equipment." *Quality and Reliability Engineering International,* Vol. 2, 1986, pp. 183–197.
12. McGuire, E. P. *Industrial Product Warranties: Policies and Practices.* The Conference Board, New York, 1980.
13. Flottman, W. W. and Worstel, M. R. "Mutual Development, Application, and Control of Supplier Warranties." *Proceedings of the Annual Reliability and Maintainability Symposium,* 1977, pp. 213–221.
14. Mamer, J. W. "Discounted and Per Unit Costs of Product Warranty." *Management Science,* Vol. 33, 1987, pp. 916–930.
15. Balaban, H. S. and Meth, M. A. "Contractor Risk Associated with Reliability Improvement Warranty." *Proceedings of the Annual Reliability and Maintainability Symposium,* 1978, pp. 123–129.
16. Menke, W. W. "Determination of Warranty Reserves." *Management Science,* Vol. 15, 1969, pp. B542–B549.
17. Balaban, H. and Retterer, B. "The Use of Warranties for Defense Avionics Procurement." *Proceedings of the Annual Reliability and Maintainability Symposium,* 1974, pp. 363–368.

CHAPTER
12

Topics in Reliability

INTRODUCTION

As discussed in earlier chapters, reliability is an important factor in engineering system designs. Its history in some ways goes back to the 1930s when probability concepts were applied to problems of electric power generation [1–3]. However, usually World War II, when Germans applied basic reliability concepts to improve reliability of their V1 and V2 missiles, is regarded as the real beginning of the reliability field [4].

During the period between 1945 and 1950, the United States Air Force, Navy, and Army performed various studies concerning failure of electronic equipment, equipment repair, and maintenance cost. As the result of their findings, the Department of Defense formed an ad hoc group in 1950 on reliability of electronic equipment. Two years later, this group became known as the Advisory Group on the Reliability of Electronic Equipment (AGREE). In 1957, the group published a report that included requirements for reliability tests, effects of storage on reliability, and minimum acceptability limits. The report played an instrumental role in setting specifications for the reliability of military electronic equipment.

Two major developments in the mathematical theory of reliability took place in the early 1950s [5]:

- In 1951, Professor W. Weibull of the Royal Institute of Technology, Stockholm, published a statistical distribution to represent the breaking strength of materials [6].

• In 1952, D. J. Davis presented failure data and the results of several goodness-of-fit tests for competing failure probability distributions [7]. This work provided the support for the assumption of exponential failure distribution which is widely used today to represent the failure behavior of various engineering items.

Also, in the early 1950s two important publications on reliability appeared: *Institute of Electrical and Electronic Engineers Transactions on Reliability* and the *Proceedings of the National Symposium on Reliability and Quality Control.*

Since the 1950s the reliability field has developed into many specialized areas, such as mechanical reliability, software reliability, human reliability, and power system reliability. Today, there are over five scientific journals fully or partially devoted to the field and more than 100 books on the subject. References 8 and 9 provide comprehensive lists of publications on reliability.

This chapter describes various fundamental aspects of reliability directly or indirectly relating to maintainability.

RELIABILITY AND MAINTAINABILITY

The objective of both reliability and maintainability is to assure that the system or equipment manufactured will be in a state of readiness for operation when required, capable of carrying out its designated functions effectively, and able to meet all the required maintenance characteristics during its life span.

Reliability

As discussed in earlier chapters, reliability refers to the likelihood that a product will be, during a given period, in adequate condition to carry out its intended functions and that it will not experience failure.

Some general principles of reliability include design that precludes or minimizes failure, that increases simplicity and the standardization of parts, that uses parts with proven reliability, and that consider human factors and safety issues [10].

Maintainability

The concept of maintainability has already been well defined. As discussed, some general principles of maintainability include lowering or eliminating altogether the need for maintenance; reducing life cycle maintenance costs; lowering the number, frequency, and complexity of required maintenance tasks; establishing the extent of preventive maintenance to be performed; reducing the mean time to repair; and providing for maximum interchangeability [10].

BATHTUB HAZARD RATE CONCEPT

The bathtub hazard rate curve shown in Figure 12-1 is often used to describe failure behavior of many engineering items. Its name comes from the hazard rate's resemblance to the shape of a bathtub. For the purpose of performing various reliability studies, the bathtub hazard rate curve is divided into three regions: decreasing hazard rate region, constant hazard rate region, and increasing hazard rate region [11–13].

The decreasing hazard rate region, in which the hazard rate decreases with time, is also referred to as the infant mortality region, debugging region, or burn-in period. From the manufacturer and consumer perspectives, this region results in unnecessary repair costs and interruption of product usage. Increasing the burn-in period prior to shipment, making improvements in the manufacturing process, and improving quality control activities can all minimize the occurrence of early failures. Some of the reasons for failures in this region include substandard workmanship and parts, poor manufacturing methods, human error, inadequate quality control, and unsatisfactory debugging.

The constant hazard rate region is also known as the useful life period of a product. This region begins at the end of the decreasing hazard rate region and terminates at the start of the increasing hazard rate period. As it can be seen from Figure 12-1, the hazard rate remains fairly constant over time during this region; thus the times to failure occurring during the useful life period may be described by an exponential distribution. In real life, the exponential distribution is normally used to represent the failure behavior of electronic parts as they exhibit a fairly long period of useful life. It is assumed that the failures occur randomly in the useful life region. Some of the

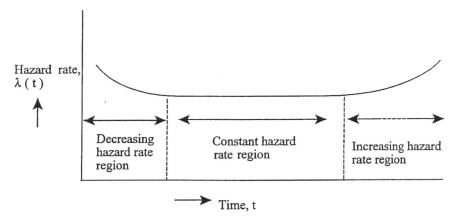

Figure 12-1. Bathtub hazard rate curve.

reasons for such failures are undetectable defects, low safety factors, high unexpected random stress, abuse, and natural failures.

The increasing hazard rate region, in which the hazard rate increases with time, is also known as the wear-out period. It begins at the end of the useful life period of the item. The failures occurring in this region are no longer random and their causes include aging, friction, wrong overhaul practices, poor maintenance, and corrosion.

RELIABILITY TERMS, DEFINITIONS, AND FORMULAS

Some of the important reliability related terms and definitions are [11]:

- **Reliability.** This is the probability that an item will carry out its stated function adequately for the specified time interval when operated according to the designed conditions.
- **Failure.** This is the inability of an item to function within the specified guidelines.
- **Hazard rate.** This is the rate of change in the number of failed items divided by the number of items that have not failed at time t.

- **Active redundancy.** This term indicates that all redundant items are functioning simultaneously.

The following are some basic formulas related to reliability.

Cumulative Distribution Function

This is defined by

$$F(t) = \int_0^t f(t)dt \tag{12.1}$$

where F(t) is the cumulative distribution function or the failure probability at time t.

   f(t) is the probability density function or failure density function.

Reliability Function

Subtracting Equation 12.1 from unity, we get the reliability function

$$R(t) = 1 - F(t) = 1 - \int_0^t f(t)dt \tag{12.2}$$

Alternatively, the reliability function is defined as:

$$R(t) = \int_t^\infty f(t)dt \tag{12.3}$$

or

$$R(t) = e^{-\int_0^t \lambda(t)dt} \tag{12.4}$$

where $\lambda(t)$ is known as hard rate or instantaneous failure rate.

Note that Equations 12.2 through 12.4 yield the same result.

Hazard Rate

This is expressed by

$$\lambda(t) = \frac{f(t)}{R(t)} \tag{12.5}$$

or

$$\lambda(t) = \frac{f(t)}{1 - F(t)} \tag{12.6}$$

Mean Time to Failure (MTTF)

This can be defined in three different ways:

- $$MTTF = \int_0^\infty tf(t)dt \tag{12.7}$$

- $$MTTF = \int_0^\infty R(t)dt \tag{12.8}$$

- $$MTTF = \lim_{s \to 0} R(s) \tag{12.9}$$

where $R(t)$ is the reliability at time t.
        s is the Laplace transform variable.
        $R(s)$ is the Laplace transform of the reliability function.

Example 12-1

Analysis of the accumulated failure data for an electronic device established that the times to failure of the device can be described by the following probability density function:

$$f(t) = \lambda e^{-\lambda t} \tag{12.10}$$

where λ is the distribution parameter or the constant failure rate of the electronic device.

Obtain expressions for the device's reliability, the device's hazard rate, and the device's mean time to failure.

Inserting Equation 12.10 into Equation 12.2 yields

$$R(t) = 1 - \int_0^t \lambda e^{-\lambda t} dt = e^{-\lambda t} \tag{12.11}$$

Substituting Equations 12.10 and 12.11 into Equation 12.5, we get

$$\lambda(t) = \frac{\lambda e^{-\lambda t}}{e^{-\lambda t}} = \lambda \tag{12.12}$$

By inserting Equation 12.11 into Equation 12.8 we obtain

$$MTTF = \int_0^\infty e^{-\lambda t} dt = \frac{1}{\lambda} \tag{12.13}$$

Thus, the electronic device's reliability, hazard rate, and mean time to failure are given by Equations 12.11, 12.12, and 12.13, respectively.

Example 12-2

Prove by using Equations 12.9 and 12.11 that the end result for the electronic device's mean time to failure is same as given by Equation 12.13.

The Laplace transform of a time function, f(t) is defined by [14]

$$F(s) = \int_0^\infty f(t) e^{-st} dt \tag{12.14}$$

where F(s) is the Laplace transform of f(t).

Substituting Equation 12.11 into Equation 12.14 yields

$$F(s) = \int_0^\infty e^{-\lambda t} e^{-st} dt = \int_0^\infty e^{-(s+\lambda)t} dt = \frac{e^{-(s+\lambda)t}}{(s+\lambda)} \bigg|_0^\infty = \frac{1}{s+\lambda} \tag{12.15}$$

Thus, the Laplace transform of the electronic device's reliability function is

$$R(s) = F(s) = \frac{1}{s + \lambda} \qquad (12.16)$$

Inserting Equation 12.16 into Equation 12.9, we get

$$MTTF = \lim_{s \to 0} \frac{1}{(s + \lambda)} = \frac{1}{\lambda} \qquad (12.17)$$

The result for the device's mean time to failure given by Equation 12.17 is exactly the same as given by Equation 12.13. It proves that Equations 12.8 and 12.9 yield identical results.

STATIC STRUCTURES

These are those structures or networks whose unit reliability or failure probability does not change with time. In other words, the reliability remains constant. The basic static structures used in reliability and maintainability work include series, parallel, and r-out-of-m units.

Series Structure

This is the simplest and perhaps the most commonly used structure in reliability studies. In the series structure, it is assumed that all its elements or units, as shown in Figure 12-2, are connected in series. If any one of the units fails, the structure fails. More specifically, all the structure units must function normally for the successful operation

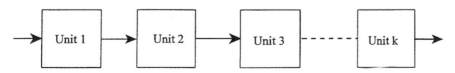

Figure 12-2. A k unit series structure.

of the structure. If the events x_1, x_2, x_3, . . . , x_k denote the success of units 1, 2, 3, . . . , k, respectively, as shown in Figure 12-2, the probability of the series structure success is given by

$$P_s = R_s = P(x_1, x_2, x_3 \ldots x_k) \tag{12.18}$$

where R_s is the series structure reliability.
$P(x_1, x_2, x_3 \ldots x_k)$ is the probability of occurrence of success events x_1, x_2, x_3, and x_k.

For independent events or unit failures, Equation 12.18 becomes

$$R_s = P(x_1)P(x_2)P(x_3) \ldots P(x_k) = R_1R_2R_3 \ldots R_k \tag{12.19}$$

where $P(x_i) \equiv R_i$, for i = 1, 2, 3, . . . , k.
$P(x_i)$ is the probability of occurrence of success event i, for i = 1, 2, 3, . . . , k.
R_i is the reliability of unit i, for i = 1, 2, 3, . . . , k.

The series structure unreliability is given by

$$F_s = 1 - R_s = 1 - R_1R_2R_3 \ldots R_k \tag{12.20}$$

Since the unreliability or failure probability of unit i plus its reliability is always equal to unity, the reliability of unit i can be expressed as

$$R_i = 1 - F_i \tag{12.21}$$

where F_1 is the failure probability of unit i, for i = 1, 2, 3, . . . , k.

Substituting Equation 12.21 into Equations 12.19 and 12.20, we get

$$R_s = (1 - F_1)(1 - F_2)(1 - F_3) \ldots (1 - F_k) \tag{12.22}$$

and

$$F_s = 1 - (1 - F_1)(1 - F_2)(1 - F_3) \ldots (1 - F_k) \tag{12.23}$$

Example 12-3

Assume that an aircraft has four independent and identical engines and that all of them must operate normally for the aircraft to fly successfully. If the reliability of each engine is 0.98, calculate the probability of the aircraft flying successfully.

In this case we have k = 4, and $R = R_1 = R_2 = R_3 = R_4 = 0.98$. Substituting these specified values into Equation 12.19 yields

$$R_s = (0.98)(0.98)(0.98)(0.98) = 0.9224$$

Thus the probability of the aircraft flying successfully is 0.9224.

Parallel Structure

In this case, the structure consists of k active units and at least one of the k units must function normally for system success. The system fails only when all of the k units fails. Figure 12-3 shows a block

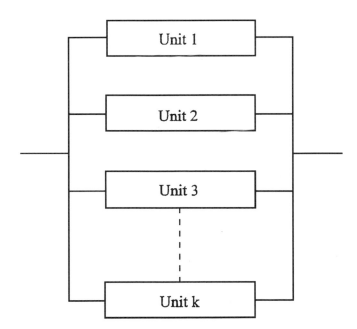

Figure 12-3. A parallel structure.

diagram representing a parallel structure. Note that all the units in this configuration are active and load sharing.

If the events $\overline{x_1}, \overline{x_2}, \overline{x_3} \ldots, \overline{x_k}$ denote the failure of units 1, 2, 3, ..., k, respectively, as shown in Figure 12-3, the probability of the parallel structure failure is expressed by

$$P_{pf} = F_p = P(\overline{x_1}\,\overline{x_2}\,\overline{x_3}\,\overline{x_k}) \qquad (12.24)$$

where F_p is the failure probability of the parallel structure.
$P(\overline{x_1}\,\overline{x_2}\,\overline{x_3}\,\overline{x_k})$ is the probability of occurrence of failure events $\overline{x_1}$, $\overline{x_2}, \overline{x_3}$, and $\overline{x_k}$.

If all the units fail independently, Equation 12.24 can be written to the following form:

$$F_p = P(\overline{x_1})P(\overline{x_2})P(\overline{x_3}) \ldots P(\overline{x_k}) = F_1 F_2 F_3 \ldots F_k \qquad (12.25)$$

where $P(\overline{x_i}) = F_i$, for i = 1, 2, 3, . . . , k.
$P(\overline{x_i})$ is the probability of occurrence of failure event i, for i = 1, 2, 3, . . . , k.
F_i is the failure probability or unreliability of unit i, for i = 1, 2, 3, . . . , k.

Subtracting Equation 12.25 from unity, we get the following expression for the parallel system reliability

$$R_p = 1 - F_1 F_2 F_3 \ldots F_k \qquad (12.26)$$

where R_p is the parallel system reliability.

Rearranging Equation 12.21 and substituting it into Equation 12.26 yields

$$R_p = 1 - (1 - R_1)(1 - R_2)(1 - R_3) \ldots (1 - R_k) \qquad (12.27)$$

For identical units, Equation 12.27 simplifies to

$$R_p = 1 - (1 - R)^k \qquad (12.28)$$

where $R = R_i$, for $i = 1, 2, 3, \ldots, k$.

For given R_p and known R, the number of units to be placed in parallel can be obtained by rearranging Equation 12.28, for example,

$$k = \frac{\ln(1 - R_p)}{\ln(1 - R)} \qquad (12.29)$$

Example 12-4

An aircraft has two active, identical, and independent engines. At least one of the engines must operate normally for the aircraft to fly successfully. Calculate the probability of the aircraft flying successfully, if the probability of failure of an engine is 0.02.

Since the reliability plus failure probability of an engine is equal to unity, we subtract the engine failure probability from unity to get engine reliability

$$R = 1 - 0.02 = 0.98$$

Substituting the values given into Equation 12.28 yields

$$R_p = 1 - (1 - 0.98)^2 = 0.9996$$

Thus the probability of the aircraft flying successfully is 0.9996.

Example 12-5

Reliability prediction studies showed the reliability of a subsystem to be only 0.64. However, per design specification, the required reliability of that subsystem is 0.97. Consequently, the decision was made to increase the subsystem reliability through parallel redundancy. Calculate the number of independent and identical subsystems that must be connected in parallel to meet the required reliability.

Substituting the given data into Equation 12.29, we get

$$k = \frac{\ln(1 - 0.97)}{\ln(1 - 0.64)} = 3.43$$

In order to meet the specified reliability target of 0.97, there must be at least three units in parallel.

r-out-of-m Structure

In this case, a total of m active units are connected in parallel, out of which at least r units must function normally for system success. An example of this type of arrangement could be an aircraft with four active engines, out of which three must operate normally for the aircraft to fly successfully.

This structure reduces to parallel and series structures at r = 1 and r = m, respectively. For independent and identical units, using the binomial distribution, the r-out-of-m structure reliability is

$$R_{r/m} \sum_{i=r}^{m} \binom{m}{i} R^i (1-R)^{m-i} \tag{12.30}$$

where $R_{r/m}$ is the r-out-of-m structure reliability.

$$\binom{m}{i} \equiv \frac{m!}{i!(m-i)!} \tag{12.31}$$

R is the unit reliability.

Example 12-6

An engineering system consists of three active, independent, and identical subsystems. At least two subsystems must function normally for the successful operation of the system. If a subsystem's reliability is 0.9, calculate the probability of the system operating successfully.

Using Equations 12.30 and 12.31, we get

$$R_{2/3} = \sum_{i=2}^{3} \binom{3}{i} R^i (1-R)^{3-i} \tag{12.32}$$

$$\binom{3}{i} = \frac{3!}{i!(3-i)!} \tag{12.33}$$

Simplifying Equation 12.32 results in

$$R_{2/3} = \left[\frac{3!}{2!1!}\right]R^2(1-R) + \left[\frac{3!}{3!0!}\right]R^3 = 3R^2 - 2R^3 \qquad (12.34)$$

Since R = 0.9, Equation 12.34 yields

$$R_{2/3} = 0.9720$$

Thus the probability of the system operating successfully is 0.9720.

DYNAMIC STRUCTURES

In these structures, the unit reliability is a function of time. In other words, dynamic structures are basically the same as static structures except that their unit reliability varies with time. The reason for the time-dependent unit reliability is the representation of the unit times to failure by exponential, Rayleigh, Weibull, normal, and other statistical distributions.

Some of the basic dynamic structures used in reliability and maintainability work are series, parallel, r-out-or-m, and standby.

Series Structure

This structure is basically the same as the static series structure except that the times to unit failures are described by one or more probability distributions.

Thus in this case, for independent units the series system reliability from Equation 12.19 is

$$R_s(t) = R_1(t)R_2(t)R_3(t) \ldots R_k(t) \qquad (12.35)$$

where $R_s(t)$ is the series system reliability at time t.

$R_i(t)$ is the reliability of unit i at time t, for i = 1, 2, 3, . . . , k.

For exponentially distributed times to failure of each unit, the unit i reliability from Equation 12.11 is

$$R_i(t) = e^{-\lambda_i t} \tag{12.36}$$

where λ_i is the constant failure rate of unit i.

Inserting Equation 12.36 into Equation 12.35 yields

$$R_s(t) = \prod_{i=1}^{k} e^{-\lambda_i t} \tag{12.37}$$

Inserting Equation 12.37 into Equation 12.8 yields the following expression for the series system mean time to failure:

$$MTTF_s = \int_0^{\infty} \prod_{i=1}^{k} e^{-\lambda_i t} dt = \frac{1}{\displaystyle\sum_{i=1}^{k} \lambda_i} \tag{12.38}$$

where $MTTF_s$ is the series system mean time to failure.

The hazard rate of an item may be defined by

$$\lambda(t) = -\frac{1}{R(t)} \frac{dR(t)}{dt} \tag{12.39}$$

where $\lambda(t)$ is the item hazard rate or time dependent failure rate.
    R(t) is the item reliability at time t.

Inserting Equation 12.37 into Equation 12.39 yields the following expression for the series system hazard rate:

$$\lambda(t) = \sum_{i=1}^{k} \lambda_i \tag{12.40}$$

It means that whenever we add units' failure rates, we automatically assume the system units are acting in series. In other words, if any one of the unit fails, the system fails. This is the worst case design scenario.

Example 12-7

Assume that a four-wheel automobile's independent and identical tires have a failure rate of 0.0001 failures per hour. The times to failures of the tires are exponentially distributed. Calculate the reliability of the automobile with respect to tires over 12 hours of operation.

Substituting the given data into Equation 12.37 yields

$$R_s(12) = [e^{-(0.0001)(12)}]^4 = 0.9952$$

The reliability is 0.9952.

Parallel Structure

This structure was described earlier for the static case. As with the series time-dependent structure, for independent units from Equation 12.26, the parallel structure reliability is

$$R_p(t) = 1 - F_1(t)F_2(t)F_3(t) \ldots F_k(t) \tag{12.41}$$

where $R_p(t)$ is the parallel system reliability at time t.
  $F_i(t)$ is the failure probability or unreliability of unit i at time t, for $i = 1, 2, 3, \ldots, k$.

For exponentially distributed times to failure of each unit, subtracting Equation 12.36 from unity, the unit i failure probability at time t is

$$F_i(t) = 1 - e^{-\lambda_i t} \tag{12.42}$$

Substituting Equation 12.42 into Equation 12.41 yields

$$R_p(t) = 1 - \prod_{i=1}^{k} (1 - e^{-\lambda_i t}) \tag{12.43}$$

For identical units, Equation 12.43 simplifies to

$$R_p(t) = 1 - (1 - e^{-\lambda t})^k \tag{12.44}$$

where λ is the constant failure rate of a unit.

Inserting Equation 12.44 into Equation 12.8 yields

$$\text{MTTF}_p = \int_0^\infty [1 - (1 - e^{-\lambda t})^k]dt = \frac{1}{\lambda} \sum_{i=1}^k \frac{1}{i} \tag{12.45}$$

where MTTF_p is the parallel system mean time to failure.

Example 12-8

A computer has two independent and identical central processing units (CPUs) working simultaneously in parallel. The failure rate of a CPU is 0.005 failures per hour. Calculate the mean time to failure of the parallel system with two CPUs.

Substituting the given data into Equation 12.45 yields

$$\text{MTTF}_p = \frac{1}{(0.005)} \sum_{i=1}^2 \frac{1}{i} = 300 \text{ hours} \tag{12.45}$$

Thus the mean time to failure of the parallel system with two CPUs is 300 hours.

r-out-of-m Structure

This structure was described earlier as a static structure. As is the case of series and parallel structures, the time-dependent reliability of the r-out-of-m structure with independent and identical units and exponentially distributed unit failure times is

$$R_{r/m}(t) = \sum_{i=r}^m \binom{m}{i} e^{-i\lambda t}(1 - e^{-\lambda t})^{m-i} \tag{12.46}$$

where $R_{r/m}(t)$ is the reliability of the r-out-of-m structure at time t. λ is the unit constant failure rate.

Inserting Equation 12.46 into Equation 12.8 yields

$$\text{MTTF}_{r/m} = \int_0^\infty \left[\sum_{i=r}^m \binom{m}{i} e^{-i\lambda t}(1 - e^{-\lambda t})^{m-i} \right] dt = \frac{1}{\lambda} \sum_{i=r}^k \frac{1}{i} \qquad (12.47)$$

where $\text{MTTF}_{r/m}$ is the r-out-of-m structure mean time to failure.

Example 12-9

An aircraft has four independent and identical active engines, and at least three must operate normally for the aircraft to fly successfully. Calculate the probability of the success of an 11-hour mission, if the failure rate of each engine is 0.0006 failures per hour.

Using the specified values in Equation 12.46, the aircraft's reliability is

$$R_{3/4}(11) = \sum_{i=3}^4 \binom{4}{i} e^{-i(0.0006)(11)}[1 - e^{-(0.0006)(11)}]^{4-i}$$

$$= 4e^{-3(0.0006)(11)} - 3e^{-4(0.0006)(11)} = 0.9997$$

Thus the probability of success is 0.9997.

Standby Structure

The standby structure represents a form of redundancy in which only one unit operates and k units remain in their standby mode. Whenever the operating unit fails, it is immediately replaced by one of the standby units. The standby structure fails only when the operating unit and all the standby units fail. Figure 12-4 is a block diagram representing the standby structure.

For (k + 1) independent and identical units with exponentially distributed times to failures, the reliability of the standby structure is [11]

$$R_{SB}(t) = e^{-\lambda t} \sum_{i=0}^k \frac{(\lambda t)^i}{i!} \qquad (12.48)$$

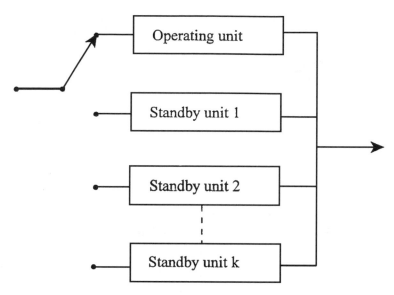

Figure 12-4. Block diagram of a (k + 1) unit standby system.

where $R_{SB}(t)$ is the reliability of the standby structure at time t.
λ is the constant failure rate of a unit.

Note that Equation 12.48 involves the following two assumptions:

- The switching mechanism to replace the failed unit with a good one never fails.
- The standby units remain as good as new in their standby mode.

Substituting Equation 12.48 into Equation 12.8, we get the following expression for the standby structure mean time to failure:

$$\text{MTTF}_{SB} = \int_0^\infty \left[e^{-\lambda t} \sum_{i=0}^{k} \frac{(\lambda t)^i}{i!} \right] dt = \frac{k+1}{\lambda} \tag{12.49}$$

where MTTF_{SB} is the standby structure mean time to failure.

Example 12-10

Assume that a system has two independent and identical pumps operating in the standby structure arrangement—only one pump operates and the other remains in its standby mode. Calculate the reliability of the standby structure for a 10-hour mission, if the pump failure rate is 0.004 failures per hour and the switching mechanism to replace the failed pump is perfect. Use Equation 12.48 to obtain the end result.

Substituting the specified data into Equation 12.48 yields

$$R_{SB}(10) = e^{-(0.004)(10)} \sum_{i=0}^{1} \frac{[(0.004)(10)]^i}{i!}$$

$$= e^{-(0.004)(10)}[1 + (0.004)(10)] = 0.9992$$

The reliability of the pump standby structure is 0.9992.

SYSTEM AVAILABILITY

System or item availability is one of the most important measures of maintained systems or items since it takes into consideration both reliability and maintainability. More specifically, availability takes into account both failure and repairability of a system. This section discusses three different types of availability: instantaneous availability, steady state availability, and average up-time availability.

Instantaneous and Steady State Availabilities

A problem concerning failure and repair of a system can demonstrate the concept of instantaneous availability. For constant failure and repair rate, the Markov technique can give the instantaneous or time-dependent availability of a single item or system [14]. Figure 12-5 presents the state space diagram of a single repairable system. The numerals in the box and in the triangle of Figure 12-5 denote system state. The system represented by the diagram can either be in operating or failed state. The numeral 0 denotes the system operating state and the numeral 1 denotes the system failed state.

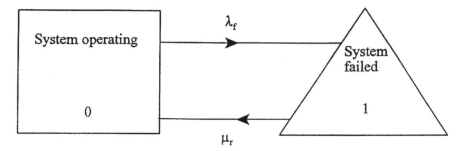

Figure 12-5. System state space diagram.

The Markov technique is based on the following assumptions:

- The probability of transition from one state to another in a finite time interval t is given by λ_f t, as in the case of Figure 12-5, where λ_f is the constant failure rate from the system operating state to the system failed state. Similarly, the probability of transition from system state 1 to system state 0 in finite time t is given by μ_r t, where μ_r is the system constant repair rate.
- The probability of more than one transition from one state to another in a finite time interval t is negligible.
- All the occurrences are independent of each other.
- System failure and repair times are exponentially distributed.

Using the Markov technique and Figure 12-5 we write the following equations:

$$P_0(t + t) = P_0(t)(1 - \lambda_f t) + P_1(t)\mu_r t \qquad (12.50)$$

$$P_1(t + t) = P_1(t)(1 - \mu_r t) + P_0(t)\lambda_f t \qquad (12.51)$$

where λ_f is the system constant failure rate.
μ_r is the system constant repair rate.
$P_0(t)$ is the probability that the system is in operating state 0 at time t.
$P_1(t)$ is the probability that the system is in failed state 1 at time t.
$(1 - \lambda_f t)$ is the probability of no failure in time interval t when the system is in state 0.

$P_0(t + t)$ is the probability of the system being in operating state 0 at time $t + t$.

$(1 - \mu_f t)$ is the probability of no repair in time interval t when the system is in state 1.

$P_1(t + t)$ is the probability of the system being in failed state 1 at time $t + t$.

$\lambda_f t$ is the probability of system failure in time interval t.

$\mu_f t$ is the probability of accomplishing system repair in time interval t.

In the limiting case Equations 12.50 and 12.51 become

$$\lim_{\Delta t \to 0} \frac{P_0(t + \Delta t) - P_0(t)}{\Delta t} = \frac{dP_0(t)}{dt} = P_1(t)\mu_r - P_0(t)\lambda_f \qquad (12.52)$$

$$\lim_{\Delta t \to 0} \frac{P_1(t + \Delta t) - P_0(t)}{\Delta t} = \frac{dP_1(t)}{dt} = P_0(t)\lambda_f - P_1(t)\mu_r \qquad (12.53)$$

At time $t = 0$, $P_0(0) = 1$ and $P_1(0) = 0$.
Solving Equations 12.52 and 12.53, we get [11]

$$P_0(t) = \frac{\mu_r}{\lambda_f + \mu_r} + \frac{\lambda_f}{\lambda_f + \mu_r} e^{-(\lambda_f + \mu_r)t} \qquad (12.54)$$

and

$$P_1(t) = \frac{\lambda_f}{\lambda_f + \mu_r} - \frac{\lambda_f}{\lambda_f + \mu_r} e^{-(\lambda_f + \mu_r)t} \qquad (12.55)$$

Thus, the system's instantaneous availability is given by

$$AV(t) = P_0(t) = \frac{\mu_r}{\lambda_f + \mu_r} + \frac{\lambda_f}{\lambda_f + \mu_r} e^{-(\lambda_f + \mu_r)t} \qquad (12.56)$$

where AV(t) is the system's instantaneous availability. It can be obtained for any specified time t.

The system steady state availability is expressed by

$$AV = \lim_{t \to \infty} AV(t) \tag{12.57}$$

For our case, we substitute Equation 12.56 into Equation 12.57 to get

$$AV = \frac{\mu_r}{\lambda_f + \mu_r} \tag{12.58}$$

This equation gives the steady state availability of a single repairable system.

Example 12-11

Assume that the constant failure and repair rates of a computer are 0.0001 failures per hour and 0.0005 repairs per hour, respectively. Calculate the computer's steady state availability.

Inserting the specified data into Equation 12.58 yields

$$AV = \frac{0.005}{0.0001 + 0.005} = 0.9804$$

Thus the steady state availability of the computer is 0.9804.

Average Up-Time Availability

This is an important measure whenever it is desirable to state availability requirements with respect to the proportion of time in a given time period, say 0 to T, that the system or item is available for service. This availability is probably the most useful measure for systems whose usage is defined by a duty cycle. One example is a tracking radar system [15]. The average up-time availability is defined by

$$AV_a(T) = \frac{1}{T} \int_0^T AV(t) dt \tag{12.59}$$

where $AV_a(T)$ is the average up-time availability of a system or item. $AV(t)$ is the instantaneous availability of a system or item.

RELIABILITY DATA SOURCES

The availability of reliability data is an important component in reliability studies. Without good failure data, the studies can lead to incorrect results concerning the reliability of products or items. Over the years, there has been significant effort to collect and analyze various types of failure data. Some important sources for obtaining reliability data are:

* MIL-HDBK-217, *Reliability Prediction of Electronic Equipment.* Department of Defense, Washington, D.C. This is probably the most widely used document to compute failure rates of electronic parts.
* *NERC Data,* The National Electric Reliability Council (NERC), New York. NERC publishes failure data collected from the United States power plants annually.
* GIDEP Data. The Government Industry Data Exchange Program (GIDEP) is a computerized data bank managed by the GIDEP Operations Center, Fleet Missile Systems, Analysis and Evaluation Group, Department of Defense, Corona, California.
* FARADA (Failure Rate Data). These are reports containing part failure probabilities derived from Department of Defense and National Aeronautics and Space Administration experience. The reports are released by the Fleet Missile Systems, Analysis and Evaluation Group, Department of Defense, Corona, California.

PROBLEMS

1. Define the following terms:
   * Reliability
   * Maintainability
   * Availability
2. Discuss the relationship between product reliability and maintainability.

3. Write an essay on reliability engineering by emphasizing its history.
4. Write down the probability density function of a distribution that can represent the bathtub hazard rate curve.
5. Prove that the reliability of an item is given by

$$R(t) = e^{-\int_0^t \lambda(t)dt}$$

where $R(t)$ is the item reliability at time t.
  $\lambda(t)$ is the hazard rate of the item.
6. Prove that the meantime to failure (MTTF) of an item is expressed by

$$MTTF = \lim_{s \to 0} R(s)$$

where s is the Laplace transform variable.
  $R(s)$ is the Laplace transform of the item reliability function, $R(t)$.
7. Prove that the mean time to failure of a system is given by

$$MTTF = \frac{1}{\lambda_1} + \frac{1}{\lambda_2} - \frac{1}{(\lambda_1 + \lambda_2)}$$

where λ_1 is the constant failure rate of unit 1.
  λ_2 is the constant failure rate of unit 2.
8. An aircraft has three independent and identical active engines. At least two engines must work normally for the aircraft to fly successfully. If the failure rate of an engine is 0.0007 failures per hour, calculate the probability of the aircraft flying successfully for 8 hours. State any assumptions made in performing your calculations.
9. Prove using the Markov method that the reliability of a two-unit standby system (one unit operating, the other on standby) is given by

$$R_{SB}(t) = e^{-\lambda t}(1 + \lambda t)$$

where λ is the constant failure rate of a unit.
  t is time.

State any assumptions associated with your proof.
10. Write an essay on the importance of failure data.

REFERENCES

1. Layman, W. J. "Fundamental Considerations in Preparing a Master System Plan." *Electrical World,* Vol. 101, 1933, pp. 778–792.
2. Smith, S. A. "Service Reliability Measured by Probabilities of Outage." *Electrical World,* Vol. 103, 1934, pp. 371–374.
3. Benner, P. E. "The Use of the Theory of Probability to Determine Space Capacity." *General Electric Review,* Vol. 37, 1934, pp. 345–348.
4. Dhillon, B. S. *Reliability Engineering in Systems Design and Operation.* Van Nostrand Reinhold Company, New York, 1983.
5. Barlow, R. E. "Mathematical Theory of Reliability: A Historical Perspective." *IEEE Transactions on Reliability,* Vol. 33, 1984, pp. 16–20.
6. Weibull, W. "A Statistical Distribution Function of Wide Applicability," *Journal of Applied Mechanics,* Vol. 18, 1951, pp. 293–297.
7. Davis, D. J. "An Analysis of Some Failure Data." *Journal of American Statistical Association,* Vol. 47, 1952, pp. 113–150.
8. Dhillon, B. S. *Reliability and Quality Control: Bibliography on General and Specialized Areas.* Beta Publishers, Inc., Gloucester, Ontario, 1992.
9. Dhillon, B. S. *Reliability Engineering Applications: Bibliography on Important Application Areas.* Beta Publishers, Inc., Gloucester, Ontario, 1992.
10. AMCP 706-134, *Maintainability Guide for Design.* Department of Defense, Washington, D.C., 1972.
11. Dhillon, B. S. *Engineering Design: A Modern Approach.* Richard D. Irwin, Inc., Chicago, 1996.
12. Elsayed, E. A. *Reliability Engineering.* Addison Wesley Longman, Inc., Reading, Massachusetts, 1996.
13. Ramakumar, R. *Engineering Reliability: Fundamentals and Applications.* Prentice-Hall, Inc., Englewood Cliffs, New Jersey, 1993.
14. Shooman, M. L. *Probabilistic Reliability: An Engineering Approach.* McGraw-Hill Book Company, New York, 1968.
15. Lie, C. H., Hwang, C. L. and Tillman, F. A. "Availability of Maintained Systems: A State-of-the-Art Survey." *AIIE Transactions,* Vol. 9, 1977, pp. 247–259.

Index

B. S. Dhillon, Ph.D., is professor of mechanical engineering at the University of Ottawa. He has published more than 250 articles and books on maintainability and reliability engineering, as well as on related topics. Dr. Dhillon has been teaching reliability and maintainability for over 18 years at the University of Ottawa and has received numerous awards for his work in these fields.